BEHAVIORAL GENETICS

E 70

Staff of Research and Education Association

 Research and Education Association
505 Eighth Avenue
New York, N. Y. 10018

BEHAVIORAL GENETICS

Copyright © 1982 by Research and Education Association. All rights reserved. No part of this book may be reproduced in any form without permission of the publishers.

Printed in the United States of America

Library of Congress Catalog Number 82-80748

International Standard Book Number 0-87891-537-0

PREFACE

Behavioral genetics is a relatively new field which draws on many disciplines. The influence of genetic factors in the transmission of mental illnesses such as schizophrenia has been suggested by the use of statistical methods. It is also now clear that human brains have structures and functions which are genetically determined and which are further shaped by experience.

This reference book has been prepared in a non-technical manner, and provides an integrated presentation of the many different disciplines that contribute to our knowledge of behavioral genetics. The book is intended to aid (1) in the understanding of the relationship of the neurochemistry and neurophysiology of mental illness, (2) in defining the pathways that lead from genetic endowment to overt behavior, and (3) in developing more powerful genetic strategies in biological psychiatry.

The results of various studies and research efforts are analyzed in relating the complex characteristics of nature and nurture. Also discussed, is the topic of behavior stability-- the degree of resistance or responsiveness to outside influences. Considerable use was made of investigations on twins, and correlations of adoptees and parents.

The major amount of information in this handbook was originated and sponsored by the National Insitute of Mental Health and edited by Patricia McBroom.

CONTENTS

page

SECTION I. NATURE—NURTURE

3	Chapter I. Heredity Matters
11	Chapter II. The Search Begins

SECTION II. GENETIC PATHWAYS

27	Chapter III. Known Pathways in Animals
37	Chapter IV. Potential Pathways in Man

SECTION III. MENTAL DISORDERS

59	Chapter V. Schizophrenia
85	Chapter VI. Depression
113	Chapter VII. Children at Risk

SECTION IV. PERSONALITY

129	Chapter VIII. Normal Personality
149	Chapter IX. Selected Abnormal Traits
149	Part I. Sex Chromosome Defects
157	Part II. Psychopathy

SECTION V. INTELLIGENCE

167	Chapter X. Intelligence
189	**GLOSSARY**

AFTERWORD—*Staff, NIMH*

195	Part I. Genetic Counseling
213	Part II. Directions of Future Research in Behavioral Genetics

APPENDIX

220	Fundamental Concepts in Behavioral Genetics

Section I.

NATURE—NURTURE

> See the Appendix on page 220 for a discussion and review of fundamental concepts in behavioral genetics.

Chapter I. Heredity Matters

In a fraction of geological time—three million years or less—nature added more than a pound of brain cells to the budding human brain of some unknown primate. As the new cells intercalated to expand and fold into the underlying brain structures, the new brain tripled in volume and allowed its possessor to tear himself loose bit by bit from the animal world.

The new brain provided language, and men began to talk and to build the base for a culture that would expand far beyond any biological limitations. It provided the capacity for increased logic and reasoning ability that could compare sensory information and thoughts, not restrained by emotional input from older parts of the animal brain. But most remarkable of all, the new brain improved the ability to learn, and, by that means, to diminish the control exercised by innate, genetically based behavior.

A NEW SELF-SUFFICIENCY

Unlike the older structures, this pliable, nearly formless material could shift functions from one location to another. Moreover, this new cortex could communicate only within itself, without referring any nerve signals through the emotionally laden animal brain beneath. Thus, intellectual processes could be carried out with relatively little input from regions involved with aggression, sexuality, eating, sleeping, and a host of other basic behaviors.

The cerebral cortex is nature's experiment with experience. It is so adaptable that removal of a chunk of cortex by a scientist from the brain of a newborn monkey apparently had no effect on the animal's normality as an adult. Whatever function that brain material had served was apparently taken over by other cortical areas. That same operation, if done on older structures beneath the cerebral cortex, would produce marked changes in behavior, from senseless aggression to strange sexual behavior and confused body functions. There is very little in the old prehuman brain that is exchangeable.

With this kind of brain function, the human can inhibit some instincts, rather than be controlled by them, to a degree unprecedented in the animal kingdom. What is left of genetic control over human behavior is what this book is about.

At birth, a human infant has only 25 percent of his future brain weight. Compared to other young primates, he is highly incompetent as a youngster, requiring years of learning and maturation before he can care for himself. The long childhood dependency is a hallmark of advanced species—a crucial development in evolution that allows extensive cultural and biological transmission between parents and offspring, in preparation for the more complex life of human beings.

THE CONCEPT OF BEHAVIOR

On the basis of this educability, one reaction might be to conclude that genes have no role left—nothing important anyway—in controlling human behavior. Of what use are genes when social habits, personal traits, and intellectual skills are learned? Indeed, since the 1930s, social philosophers and scientists have given heredity a minor role, if any, in the central operations of human behavior.

But the environmental bias of the mid-century was too extreme. The fact that genes do influence the mental and emotional lives of human beings is now being recognized. Increasingly, evidence is gathering to reveal the impact of genes on behavior, although in no way does it demonstrate that behavior is *controlled* by genes. Eye color is controlled by genes; human behavior is not. Genes make a *contribution* to sociability, aggressiveness, artistic talent, IQ, and most other human behaviors. How much of a contribution is the question.

Take a second look at that newborn brain. True, it has only 25 percent of its future brain weight. But what a 25 percent! Nearly all the weight is accounted for by neurons. Tightly packed together, all the neurons a human will have throughout life are present at birth (Rose 1973). The basic units of nerve function are laid down in the womb, under the control of genetic instructions within the limitations of the prenatal environment. How much information is programmed into those neurons? No one knows, but it is known that some things are coded. The brain is not a *tabula rasa* at birth, since it arrives with at least some predispositions.

The explosive growth after birth comes when myelin sheaths develop to protect the neurons; glial cells appear which support the neurons and nerve endings and pathways spread out to tie them together in a complex network. Many elements are integrated at that time, but the human brain is far more complete at birth than the brain of a rat, for instance, whose neurons continue to multiply postnatally. Because of that more completed condition, the human brain is more protected against the effects of postnatal starvation than is the brain of a rat.

Like any other physical organ constructed under the direction of genes, the human brain shows individual variation. No two brains are entirely alike. There is always some difference in size or location of the various parts of the brain, a possible source of genetic individuality. Furthermore, the chemical state of the brain is also influenced by genes. New evidence has demonstrated a genetic role in the synthesis of brain chemicals responsible for nerve transmission. Thus, both their structure and their biochemistry have hereditary components, and there is no longer any question but that they affect important behaviors.

But just how significant is the genetic contribution? Does it control IQ, for instance, or predispose certain people to mental illness? Does it influence personality and, if so, to what degree? What biological pathways are used by genes to affect the mental and emotional states central to behavior?

SOURCES AND SOURCE MATERIALS

These are the questions to be addressed in these pages. It is to be noted that the material contained here reflects primarily scientific work supported by the National Institute of Mental Health. Since NIMH has at one time or another funded some 85 percent of researchers in the field of behavioral genetics, no restrictions were felt in collecting the critical information here. Similarly, other genetic research has been included whenever it appeared relevant, whether or not it was supported by NIMH. The important criteria for selection were the quality of the work and the degree to which it could throw light on the central issues.

There were many ways to present new conclusions on behavioral genetics. The subject matter, ill-defined and disconnected, sprawls across half a dozen disciplines, each structured in such a way that it cannot be compared to the others. Psychiatry has something to say about the genetics of behavior. So do anthropology, psychology, biology, and medicine in addition to the young research field of behavioral genetics.

Avoiding Statistical Mazes

The subject abounds in statistical research. One can find statistical evidence for a genetic component in every conceivable type of personality and behavioral trait. Many of the studies contradict each other, and, without biological information to connect genes to behavior at least theoretically, it is difficult to know which is right.

Furthermore, the statistical work often rests on questionable data; that is, the researchers were not able to use consistent, reliable criteria for defining the trait or disorder. Things named the same from study to study may be defined differently and may denote different phenomena. Extraversion or introversion, for instance, may differ in different personality tests.

Schizophrenia, to take a more important example, is a collection of many symptoms, probably developing from several biological causes or processes. The diagnosis of this mental condition varies from one region of the country to another, from one hospital to another, and from one psychiatrist to another. The biggest diagnostic differences, as might be expected, are between countries. Statistical models of schizophrenia have proposed that the condition is caused by any one of several modes of transmission including: (1) a single gene, possibly dominant; (2) many genes; (3) many genes and one major gene. All of these models are feasible, if schizophrenia is evidenced in several different physical conditions.

Thus, some decisions had to be made about the kind of research to include in a current review of the field. Statistics have been used sparingly. They are included primarily to indicate the risk of breakdowns in relatives of mental patients.

Heritability statistics based on twin studies that attempt to estimate the relative importance of a genetic component, as compared to environmental influence, were for the most part not used, for the reason that many scientists have raised questions about the reliability of twin studies in genetic research.

Thus, the twin research was used to indicate the presence of a genetic component (but not its size) when other methodology also supported that conclusion.

Every effort was made to report statistical research when it was accompanied by biological findings of potential genetic pathways. That choice was based on the belief that one can learn little from a study which says that genes contribute, for instance, to antisocial behavior but does not say how.

Finally, genetic models of transmission were included to indicate trends in thinking, but not in any detail. It seemed misleading to discuss schizophrenia as resulting from either a single gene or multiple genes, when the behavorial and biological processes remain unclear.

These decisions led almost inevitably to a focus on such fundamental questions as: What is schizophrenia, or depression, or personality to the people who are searching for the action of genes? How are they defining behavior; what are they looking at; what are they finding?

So this is the story of a scientific search—a 100-year-old drive to answer one of the crucial questions of history: How and to what extent does a man's heredity affect his psychological life? The material was organized to give evidence of (1) a genetic factor for specific traits, (2) the importance of that factor as compared to environmental effects, and (3) the potential pathways through which genes might be operating on behavior.

VARIABILITY OF GENETIC IMPACTS

Discussions with scientists in this field and reviews of the literature lead to the conclusion that genes affect all aspects of human behavior on a continuing basis throughout life. The concept of continuing impact is important because many think of heredity as destiny—immutable, preordained, and fixed. But while the genetic code itself is fixed, its impact on behavior is not. Personality and intelligence change constantly through the years, in adulthood as well as in childhood. Unlike stature or facial features, human behavior never becomes an achieved fact locked in place with bone and muscle. Thus, many of the concepts used in the past to describe genetic influence on physical features don't work well in the realm of behavior. A better analogy is genetic influence on the skin.

Like behavior, the skin is continually in a state of change and renewal. Day by day, and under the direction of the genetic code, old skin cells fall away and new ones are produced. As this process of replenishment continues, the genetic instructions undergo changing interpretation over time, so that subtly the skin changes color, texture, and moisture content. But the environment makes even more dramatic alterations, at times overwhelming genetic instructions. At one point, for instance, a hot desert sun may overtake and control the fate of the skin cell. At other times, a mild environment may have little or no effect, leaving the genes to exert more control. But most of the time, the qualities of the skin cell respond to both influences.

The brain and nervous system also respond day by day to the action of both genes and environment. In fact, it is more than likely that both use the same biological pathways and are physically represented in the brain through the same biochemical systems. John Fuller and W. Robert Thompson make this point in their classic 1960 work, *Behavior Genetics,* when they say that learning may be something like mutation. Like a gene, it leaves a physical modification in nerve structure. The most recent biological evidence leaves little doubt that the pathways to behavior are open to change from either direction—from the genetic code or the environment, or from the interaction between the two.

So the question becomes not whether genes affect behavior but how much.

Parents want to know what impact their rearing has on the behavior of their children. Adoptive parents, in particular, wonder whether they have any impact at all, while the parents of mentally ill offspring suffer agonies in search for basic causes.

For each individual, the feebleminded, the brilliant, the criminal, the fanatic, the wild, the shy, the depressed, the lonely, the sociable, from whatever racial, ethnic, or national group, whether writer, banker, baker—for each of us, the question arises: What shapes our behavior?

SWINGS OF EMPHASIS ON GENETICS

Different decades have given different answers. Since Charles Darwin published *The Origin of Species* in 1859, men and women have argued, often acrimoniously, over the relative importance of nature and nurture. During many early years of this century, the pendulum of scientific opinion swung heavily toward a genetic emphasis. People were "born to be criminal" or "born to be poor." Many individuals, thought to have bad genes, found themselves at the mercy of misguided laws and social customs. Because of an hereditarian-based prejudice, thousands were involuntarily sterilized, and the Nation passed laws limiting immigration of certain national groups. The culmination of this kind of bias, at its most destructive, appeared in the rationales used to justify Nazi-inspired genocide.

Then, during the second quarter of the century, leaders of social thought in this country turned dramatically against genes as an explanation for *any* human behavior. These were the years when environment and childrearing were overemphasized as explanatory variables and only a few scientists continued the search for genetic components.

Since the 1950s the pendulum has been swinging back, away from the exclusively environmental answer. Some fear it will swing too far and that society may again indulge in genetic overemphasis, which raises the threat of socially unwise policies. Others believe it has come to rest in the middle with a healthy recognition that genes and environment always work together in human behavior and that neither alone can explain the way we are.

Chapter II. The Search Begins

CHARLES DARWIN

The theory of evolution destroyed the lines man had drawn between himself and other animals. Charles Darwin's *The Origin of Species* swept like a tide through Europe, overturning within a matter of years the beliefs that supported man's vision of himself as separate from the rest of the animal kingdom.

Strangely enough, Darwin said very little about man in his work on evolution but passed over the subject with the comment ". . . much light will be thrown on the origin of man and his history."

Darwin came back to humans several years later in *Descent of Man and Selection in Relation to Sex:* "So in regard to mental qualities, their transmission is manifest in our dogs, horses, and other domestic animals. Besides special tastes and habits, general intelligence, courage, bad and good temper, etc., are certainly transmitted. With man we see similar facts in almost every family; and we now know . . . that genius, which implies a wonderfully complex combination of high faculties, tends to be inherited; and, on the other hand, it too is certain that insanity and deteriorated mental powers likewise run in the same families" (1871, pp. 106-7).

SIR FRANCIS GALTON

The first behavorial geneticist, another genius and a cousin of Darwin, truly applied the theory of evolution to human behavior. Sir Francis Galton, a wealthy man of leisurely scientific interests, was galvanized by evolutionary theory. From the moment of his "conversion," as he called it, Galton devoted himself to the study of human heredity.

Within 10 years of Darwin's historic work, Galton had published his own classic, a pedigree analysis of poets, scientists, statesmen, judges, and other outstanding men of his day. In *Hereditary Genius,* Galton showed that the families of these men would contain an unusually large number of gifted relatives. He thought this would show that ability runs genetically in families. It does run in families, of course, but Galton, like other early hereditarians, minimized the role that wealth, rearing, and family connections might have in creating a talented elite. Good genes will rise to the top of the social hierarchy, said Galton: "If a man is gifted with vast intellectual ability, eagerness to work and power of working, I cannot comprehend how such a man should be repressed" (1914, p. 35).

Many "Firsts"

Contemporary behavioral geneticists would have trouble with Galton's social conclusions, but they cannot fault his contribution to the field. He founded the field (McClearn and DeFries 1973), and his genius at scientific invention has seldom been surpassed. Galton was the first in modern history to use twins in genetic research, for somehow he knew before most people that identical twins had similar genetic components. He pioneered many of the statistical methods of analysis still in use today, and he was among the first to realize that acquired characteristics could not be inherited. These are remarkable accomplishments for a period before scientists had knowledge of genes and genetic transmission.

Galton loved to classify people. He spent generous amounts of time measuring this or that trait in thousands of individuals and then tracing those characteristics through family lines. With tools of his own invention, he tested a wide variety of physical and behavioral traits. The Galton whistle, for instance, tests hearing in the upper ranges, and with publication of *Inquiries into Human Faculty,* Galton unmasked the rich vein of individual differences in sensory discrimination, mental processes, physical features, and other traits.

Genetic Discoveries

Galton gave prizes to strangers who would send in family pedigrees through which a trait could be traced. Disease, eye color, temperament—Galton collected enough information to make basic genetic discoveries. One of his most important was the discovery that many traits show regression toward the mean. That is, the offspring of parents who are extreme on some value tend to show more moderate traits. Tall parents, for instance, will on the average have shorter children; conversely, short parents will have taller children. Although Galton did not know it, regression toward the mean is also characteristic of IQ and occurs with traits determined jointly by a large number of genes.

Galton's classifications were not confined to the laboratory. He entertained himself in the street by counting women—attractive, indifferent, repellent, and so on (Haller 1963, p. 8).

Eugenics and Breeding

Along with his scientific work, Galton did something less beneficial to mankind. He invented the word "eugenics" and started a movement to improve the human race through "good breeding." No doubt the eugenics movement would have started without Galton—the time and the state of scientific knowledge were ripe for it—but Galton provided the moral and intellectual leadership for a movement that would later seem reprehensible. At the age of 80, Galton was vigorously writing and speaking the eugenics creed with almost religious fervor. "An enthusiasm to improve the race is so noble in its aims that it might well give rise to the sense of a religious obligation," he said (1909, p. 25). Unlike other eugenists who wanted to limit the breeding of the insane, the feebleminded, the criminal, and the poor, and stop there, Galton also wanted to encourage breeding by people of "good" stock. He recommended that talented young people be given certificates of merit and financial support to marry young and breed freely (Haller 1963, p. 18).

At the turn of the century, when Galton made these proposals, eugenists were viewing men and other animals in terms of "breeding." If livestock could be selectively bred to enhance favorable qualities, then why not people? Contests were run at agricultural fairs to pick out the "fittest families," and people were tested along with the animals for mental and psychological features.

Moreover, the first eugenics group in the United States was formed in 1906 at the American Breeders Association, an eventuality that seemed perfectly logical at the time. Animal biologists were doing most of the pioneering work on the newly discovered laws of Mendelian inheritance. That year, the association established a high-powered panel to report on heredity in the human race and to propound the value of "superior blood" and the menace to society of "inferior blood." On that panel were Alexander Graham Bell, Luther Burbank, the chancellor of Stanford University, and many prominent scientists from the best universities in the country (Haller 1963, p. 62).

EUGENICS DISCREDITED

The intellectual elite who led the eugenics movement early in the century abandoned those beliefs in the 1930s, and eugenics was left blackened and discredited, not necessarily by humanitarian motives, but by scientific knowledge. Memory of certain excesses of the movement, unfortunately, has contaminated the study of behavioral genetics ever since.

It soon became clear that most human traits were not inherited as simply as the color of Mendel's garden peas. They were not determined by single genes acting in a dominant or recessive fashion, so the classic Mendelian ratio for offspring, one recessive trait to three dominants, rarely turned up in human behaviors. Multiple genes were clearly involved, and the resulting genetic picture began to look far more complex.

Secondly, advances in the biochemistry of genes revealed that traits were not inherited at all. What was inherited were chemical packets that acted to produce different traits depending on the environment at hand. An early zoologist, H.S. Jennings, gave an unusual example of this finding with the story of the Mexican salamander. In a watery world, the young grew gills, a flattened tail for swimming, and a heavy body. But in a dry environment, the same animal did not develop gills, grew small and slender, and breathed air.

Men and women do not inherit characteristics. What parents give them are chemicals which "under one set of conditions produce one set of characters; under other conditions produce another set," as Jennings observed in the salamander (1925, pp. 54–55). Moreover, the chemical code and the finished traits overlap so that each gene may contribute to many traits and each trait is influenced by the action of many genes.

By 1930, few reputable scientists would have anything more to do with eugenics, a trend reinforced by John B. Watson, founder of behaviorist psychology. Watson's special aim was to discredit heredity as a source of variation. "There is no evidence of the inheritance of traits," said Watson. The inherited structure lies ready to be shaped in a thousand different ways depending on the child's rearing.

To make his point dramatic enough, Watson issued an historic challenge that has been remembered through the decades. He said that, if he could take a dozen healthy infants and choose a special environment for them, he would guarantee to train any one of them for any pursuit in life, from doctor to thief, regardless of the child's inherent talents. "Everything called instinct is mostly a result of learning," said Watson, and succeeding generations of psychologists held to that belief, while the search for hereditary components foundered.

HEREDITY VS. REARING

A few strands of research continued—animal work and some studies of twins. A landmark study of separated twins—one of the few of its kind and still quoted—was done in 1937 by Horatio Newman, Frank Freeman, and Karl Holzinger, scientists at the University of Chicago. The team wanted to know whether identical twins, born with the same genes, would, if reared by different families, have similar IQs. Separated identical (monozygotic) twins offer a good way to distinguish heredity from home-rearing effects, but such twins are hard to find.

The Chicago group found 19 monozygotic pairs reared apart and compared them with 100 monozygotic (MZ) and dizygotic (DZ) twins reared together, a good test since fraternal dizygotic twins are like siblings and, theoretically, share half their genes on the average. Despite having different adoptive parents, the separated monozygotic twins produced mental scores more like each other than were the scores of fraternal twins who grew up in the same home (Newman, Freeman, and Holzinger 1937).

In likeness to each other, the mental scores of separated twins fell between those of identicals and fraternals reared together, an indication that the environment of a different home made some difference, but not much, in IQ level. Still, that was not the full story. Most of these twins had been adopted into homes of the same social environment from which they came. Those twins had similar IQs. But a few were adopted into very different cultural surroundings, and their scores showed considerable variation (Haller 1963, p. 172). The case histories of two pairs of these 19 sets of twins amply illustrate the strength of the genetic blueprint and the power of the cultural environment interacting with it. For example:

> Raymond and Richard were born in Indiana and separated at 1 month of age. Raymond was adopted into the home of a well-to-do physician in the city of his birth; Richard went

to the home of a truck driver in Southern Illinois. Raymond had every advantage, saw little of the seamy side of life, stayed in the same community, and had things perhaps a bit too easy. Richard's father never worked or lived very long in one place and was frequently down on his luck. Richard learned to do a great deal for himself with the help of an excellent foster mother.

Considering the differences in cultural and social advantages, these boys should have been among the most dissimilar of the 19 sets of twins. They were instead among the most similar. They looked more alike than most twins; their IQ tests were almost identical, although Raymond attended better schools while Richard was shifted about many times; both were unusually bright when tested at age 13 while in the eighth grade. There was some difference in the temperament test, Richard being more positive and more aggressive than his twin, a finding which the authors believe was due to his wider experience and the necessity for learning greater self-reliance. Otherwise there was little difference in emotional character or in nervous stability. They were as much alike as if they had been reared together.

The case is unusual for the strength of the genetic blueprint which was able to express itself under such different environments. The authors believe, however, that without Richard's foster mother, his development would have taken a different course. She shielded him from the deleterious effects of his environment without stifling him and was ambitious in his behalf.

In contrast to these boys, another case described the development of Mabel and Mary. They did take different paths, despite the fact that they were adopted by two branches of close relatives, grew up near each other in Ohio, and visited back and forth all their lives. Their separation was less complete than that of the other twins; yet they were more different from each other than most when tested at the age of 29. Mary moved with her family to town, went through high school, and took up a sedentary, indoor life clerking in her father's store and teaching piano at night. Mabel, on the other hand, became a farm woman—healthy and robust, muscular, and pounds heavier than her sister. She was educated only to the eighth grade, and was less neurotic, less emotional, and more phlegmatic.

Town-living Mary became substantially more intelligent than her twin. Separating the twins was a 17-point IQ gap, equal to 3 years of schooling, about the actual difference in their levels of education. Mary also was judged to have more nervous excitability; she was, the authors wrote, "sheltered from the cruder aspects of life and brought up definitely as a lady."

The twins had been physically very much alike at birth with no important weight or health differences, but ". . . it seems to us that the active manual labor on the farm, much of it outdoors, has had much to do with Mabel's robust health while Mary's largely sedentary, indoor life and lack of exercise have had a deleterious effect on her health and bodily vigor."

These two pairs represent extreme ends in the Newman study—one pair highly similar, the other pair different emotionally and intellectually. For the most part the rest of the separated twins showed a few differences; on the whole, they were more like each other than are siblings and fraternal twins reared together.

The twins were more alike intellectual than they were in personality, a typical finding in behavioral genetics—that mental abilities reflect a stronger, more consistent genetic component. Moreover, the IQ differences that did exist could be traced largely to differences in schooling, social class, and cultural environments.

In a few cases, the intellectual gaps between twins were as large as those found regularly among fraternal twins, a fact indicating that the environment was forcing important differences upon the same basic genetic material. When those few twin pairs were removed from the sample, differences between the other twins dwindled almost to insignificance. In other words, half a dozen twins were responsible for nearly all differences in mental ability in the sample of 19 monozygotic pairs (Newman, Freeman, and Holzinger 1937).

During these years, other scientists were looking at twins to find evidence of heredity in the major psychoses. Soon after World War II, Franz J. Kallmann published results on nearly 500 pairs of twins who exhibited schizophrenic symptoms. He claimed that nearly all of the identical pairs showed concordance for the disease, which is to say that, if one twin had the disorder, so did the other in most cases. Kallmann overestimated the degree of concordance as compared with later studies (see chapter V), but he nevertheless picked up the genetic component in that disorder when few others were paying attention.

GENETIC STUDIES ON ANIMALS

Meanwhile, through animal studies, psychologists continued to search for possible genetic components in learning ability, sensory processes, and temperament. The work leaves no doubt that mental and emotional qualities have genetic components. It also leaves no doubt of the difficulty scientists face in isolating that component. Consider the problem of intelligence in the rat.

One of the best known early scientists in this field, Robert C. Tryon, spent 20 years breeding rats for intelligence at maze-running. In several generations of selective breeding, Tryon had two distinct genetic groups whose abilities did not overlap. One group ran the maze very well and was called "bright." The other group ran the maze very poorly and was called "dull." But was this intelligence? One wonders. Further experiments indicated that the crucial genetic trait was not intelligence per se but was some reaction of the rats to that particular maze. For when they were tested on other mazes the two groups reversed themselves. The "dulls" became bright and the "brights" became dull (Fuller and Thompson 1960, p. 208). Efforts to find out what that genetic factor was led to consideration of a dozen different traits, including dislike of water, desire for food, and fear of mechanical apparatus.

Such problems have plagued the rat studies. A genetically determined behavior is isolated through selective breeding, but then enormous problems arise in trying to determine just what that behavior really is. You can't ask the rat.

John Fuller, a pioneer in behavioral genetics and selective breeding, demonstrated that point by chance 10 years ago while studying the reactions of rats to drugs. The rats had been selectively bred for avoidance learning. One breed escaped very fast from an electric shock; another breed never seemed to learn. A difference in learning ability? Not so, according to Fuller. The real difference was one of passivity versus activity. One strain cringed in fear and stayed put when it was shocked. The other strain ran.

"If you're teaching an animal to avoid, it's much easier to teach the ones whose temperament is to escape by running. It looks as though they are learning fast. Suppose you change the problem. Now the animal avoids shock by staying still. Which animal will look smarter now—the one that runs or the one whose genetic tendency is to lie down? I think that is escape behavior, not cognitive behavior."

But while the study of intelligence in a rat continues to be a problem of interpretation, much success has been achieved in breeding rats for a variety of behavioral traits—spontaneous activity, passivity, fighting behavior, susceptibility to seizures, wildness and tameness, alcohol preference and saccharine preference, to name a few. At the University of Colorado, researchers have bred two strains with striking characteristics. One strain responds to alcohol by sleeping a long time; the other responds by sleeping a short time. Study of these rats should reveal important biochemical differences relevant to alcohol use in humans.

In the most dramatic rat-breeding experiment, scientists at NIMH and Stanford were able to breed a strain for a type of fighting behavior, and, most important, the offspring showed a classic Mendelian distribution, evidence that the fighting trait was controlled by a single recessive gene (see chapter III).

Meanwhile, research on dogs was providing new insight into the genetic basis of emotions. In one 1950s experiment, 10 breeds of pet dogs were tested for fearfulness, and important strain differences were found. One group of breeds, including collies, German shepherds, poodles, corgies, and dachshunds (many hunting and working dogs), were considered "fearful" because they reacted with avoidance and wariness to various provocations such as an opening umbrella and a toy snake. Another group of dogs—boxers, Boston terriers, Bedlingtons and Scottish terriers—were called "fearless" because, instead of avoiding the provocation, they teased, inspected, or at least approached it (Mahut 1958). But many of these genetic differences were overwhelmed by special rearing. A restricted environment could raise fear levels, regardless of the animal's original disposition (Fuller and Thompson 1960, p. 258).

That does not mean, however, that the same environment causes the same effects in all animals. The predisposition channels the animal's response, so that it still shows a characteristic reaction. A toxic environment, in fact, may have exactly opposite effects on two different breeds, as D. G. Freedman demonstrated 20 years ago (McClearn and DeFries 1973, p. 167). Freedman raised four breeds in two environments. One environment was harsh; the animals were restrained and taught to obey. The second environment was indulgent, with no restrictions, no punishment.

When 7 weeks old, the animals were sent one by one into a room with a bowl of food in the center. Which pups would go to the food, after being told "no" and given a swat on the rear?

The four breeds behaved very differently. Two breeds responded in totally opposite ways. Basenjis leaped for the food as soon as they could. Shetland sheep dogs did not touch it. But really unexpected was the total lack of any rearing effect. The animals behaved according to breed lines, regardless of whether they had been raised in the harsh or the benevolent environment.

The two other breeds, by contrast, showed effects of the way they were raised. Beagles and wire-haired terriers both behaved in the same way. The animals given harsh treatment lunged for the food the minute the experimenter was out of sight, while well-treated dogs hung back much longer.

Not only were the dogs temperamentally different, the strength of the genetic factor vis-à-vis the environment varied according to breed.

If animals are genetically inclined in opposite directions, the effect of a noxious environment may be to make the animal even more like itself, more extreme in its inclinations. The stress of isolation, for instance, caused terriers to become more active than normal. They would jump and spin around during the test situation. Beagles, on the other hand, became more quiet. In both cases, animals raised as pets behaved in less extreme ways than their isolated counterparts, but they kept the same position relative to each other. Pet terriers were more active than pet beagles (McClearn and DeFries 1973, p. 170).

Fuller, drawing on years of animal research, believes that dogs can tell us about simple, basis aspects of personality. "They show many characteristics that are fundamental in human beings, social affiliation, for instance: We see a great deal of variety in the amount of bonding between animals, both as individuals and as breeds. The opposite of that—aggressive behavior—also shows genetic variation. I think the things strongly influenced by genes are aggressivity, emotionality and sociability."

Many behavioral geneticists accept the temperament theory of personality, which refers to an underlying genetic predisposition toward certain emotional, social, and activity-level traits. But the kinds of qualities called "temperament" differ from scientist to scientist. Fuller believes the concept means (1) the amount of spontaneous activity shown by an animal (activity versus passivity) (2) the strength of its curiosity or investigative behavior, and (3) its emotional reactivity or nervousness. Others would include sociability among the temperament traits.

A NEW ERA OF RESEARCH

Thus far our discussion summarizes first-generation research in behavioral genetics—the selective-breeding and early twin work. The 1950s and 1960s brought a second generation of work that split into two branches—mathematical and quantitative on the one hand, biological on the other.

Quantitative genetic research came into its heyday with the arrival of psychological testing. With personality and intelligence tests scientists hoped they could reduce these intangibles to "hard data" by finding their components, testing their appearance differentially in twins, and arriving at a firm estimate of genetic influence. It was hard work, because many of the tests tapped different personality dimensions under similar trait names, and results were all too often inconsistent. But the work provided a foundation of evidence and a sharpening of the definition of behavioral concepts.

An Important Biochemical Discovery

The biological branch, contingent on advances in the neurosciences, took longer to develop. The aim here is to find and specify the genetic pathways to behavior, the complicated chains of neural paths and chemical conversions that take place. Ultimately the aim is to treat mental conditions with biological precision based on knowledge of these genetic components. Genetic pathways to behavior are extremely difficult to find, but the rewards for doing so are very high.

Behavioral geneticists always mention one disability in particular when they talk about the promise of their field—phenylketonuria or PKU. This condition results from a recessive gene which causes an enzyme to malfunction, or not to function at all. Untreated PKU babies, although normal at birth, become rapidly retarded from the poisonous byproduct which floods the brain, and they frequently need life-long institutionalization. But if the condition is detected very early, and if the baby is placed on a special diet, much of the brain damage can be avoided. For this reason, newborns in the United States are routinely screened for PKU and, if they exhibit this genetic condition, are placed on a specially designed diet.

PKU stands as a model of expectations for behavioral geneticists, especially in regard to the schizophrenic and depressive disorders. Before the discovery in 1934 of the cause of PKU, it was classified as simple mental retardation. People with the condition were part of the vast heterogeneous population with subnormal intelligence caused by a variety of biological and cultural deficiencies.

But then a Norwegian nutritionist at the University of Oslo discovered that two feebleminded children had a peculiar trait. Their urine turned green in a solution of ferric chloride. Asbjorn Folling traced the novel color to an excess of phenylpyruvic acid, and subsequent research established that the buildup, which causes brain damage, occurs because PKU individuals lack a crucial enzyme or gene product for converting the amino acid, phenylalanine, into tyrosine. Instead, phenylalanine degrades into other, toxic products. This is a major metabolic pathway in the body. Nearly all food contains phenylalanine, a fact which made it difficult for nutritionists to find a suitable diet with which to treat PKU babies.

But they did find the diet; they did treat the newborns; and they did prevent thousands of individuals from becoming mentally retarded.

Behavioral geneticists hope to do the same thing with mental disorders. If schizophrenia and depression represent several distinct genetic conditions, as many people in the field believe, then scientists should be able to find one or more of those special causes and perhaps then find ways to salvage a large group from the amorphous mass of people now identified as mentally ill.

Behavioral genetics leads in one direction—toward the biology of mental disease, emotional behavior, and intelligence. Years of quantitative research have established a role for genes in all these behaviors. Now the task is to find the avenues through which those genes operate.

REFERENCES

Barchas, J.D.; Ciaranello, R.D.; Kessler, S.; and Hamburg, D.A. Genetic aspects of catecholamine synthesis. In: Fieve, R.R.; Rosenthal, D.; and Brill, H., eds. *Genetic Research in Psychiatry*. Baltimore: Johns Hopkins University Press, 1975.

Buss, A.H., and Plomin, R. *A Temperament Theory of Personality Development*. New York: John Wiley (Wiley Series in Behavior), 1975.

Darwin, C. *Descent of Man and Selection in Relation to Sex.* London: John Murray, 1871; New York: Appleton, 1873.

DeFries, J.C.; Weir, M.W.; and Hegman, J.P. Differential effects of prenatal maternal stress on offspring behavior in mice as a function of genotype and stress. *Journal of Comparative and Physiological Psychology,* 63(2):332–334, 1967.

Dobzhansky, T. *Genetics of the Evolutionary Process.* New York: Columbia University Press, 1970.

Dobzhansky, T. *Mankind Evolving.* New Haven: Yale University Press, 1962.

Dobzhanzky, T. On types, genotypes, and the genetic diversity in populations. In: Spuhler, J.N., ed. *Genetic Diversity and Human Behavior.* New York: Wenner-Gren Foundation for Anthropological Research, 1967.

Fuller, J.H., and Thompson, W.R. *Behavior Genetics.* New York: John Wiley, 1960.

Galton, F. The possible improvement of the human breed under existing conditions and law and sentiment. In: *Essays in Eugenics.* London: Eugenics Education Society, 1909.

Galton, F. *Hereditary Genius, an Inquiry Into Its Laws and Consequences.* 2nd ed. London: Macmillan, 1914.

Ginsburg, B., and Laughlin, W.S. The multiple bases of human adaptability and achievement: A species point of view. *Eugenics Quarterly,* 13(3):240–257.

Haller, M. *Eugenics.* New Brunswick, N.J.: Rutgers University Press, 1963.

Hirsch, J. *Behavior-Genetic Analysis.* New York: McGraw-Hill, 1967.

Jennings, H.S. *Prometheus, or Biology and the Advancement of Man.* New York: E.P. Dutton, 1925.

Laughlin, O.D., Jr., and d'Aquili, E.G. *Biogenic Structuralism.* New York: Columbia University Press, 1974.

Mahut, H. Breed difference in the dog's emotional behavior. *Canadian Journal of Psychology,* 12:35–44, 1958.

McClearn, G.E., and De Fries, J.C. *Introduction to Behavioral Genetics.* San Francisco: W.H. Freeman, 1973.

Newman, H.H.; Freeman, F.N.; and Holzinger, K.J. *Twins: A Study of Heredity and Environment.* Chicago: University of Chicago Press, 1937.

Orenberg, E.K.; Renson, J.; Elliot, G.R.; Barchas, J.D.; and Kessler, S. Genetic determination of aggressive behavior and brain cyclic AMP. *Psychopharmacology Communication,* 1(1):99–107.

Pilbeam, D. *The Ascent of Man.* New York: Macmillan, 1972.

Rose, S. *The Conscious Brain.* New York: Alfred Knopf, 1973.

Rosenthal, D. *Genetics of Psychopathology.* New York: McGraw-Hill, 1971.

Slater, E., and Cowie, V. *The Genetics of Mental Disorders.* London: Oxford University Press, 1971.

Section II.

GENETIC PATHWAYS

Chapter III. Known Pathways in Animals

GENETIC PATHWAYS IN A SONG

Some time during its first week of adult life, with no encouragement other than its own impulses, the male cricket begins its night song. Its folded wings open, and as they move, the edge of one wing—which carries a hard, nail-like scraper—bumps across the underside of the other, where a series of ridged teeth have grown. Back and forth, in a characteristic rhythm, the wings open and shut, rippling across each other, scraper against ridges, until the wings are set to vibrating at a frequency 10 times faster than that of the tuning string on a violin.

The song of the male is unique to its species. It must be unique since several species may inhabit the same clump of reeds, and the female who is out there in the dark summer night listening has to find her way to the proper mate. When she hears a particular rhythm, a special tempo of chirps and trills, she knows without ever having learned it that this is the calling song of her species. She follows the song to its source in the male burrow where a new pattern of chirps, the courtship song, begins. After the insects have copulated the song changes for yet a third time, and the male cricket sings what the French call a "triumphal" song, which may serve to keep the partners close by.

Each of these songs seems to be completely controlled by genes, with no learning involved. If two different species are bred to each other, the offspring will sing a new song, a version intermediate between that of the two parents, in much the same way that dark-skinned and light-skinned humans may together produce brown-skinned babies.

Moreover, females among the new hybrids will respond to the songs of their brothers rather than those of their parents, a fact that indicates that the male song and female recognition system are passed simultaneously and may even originate in the same set of genes.

Nor is the song triggered by anything in the environment. If the peripheral nerves of the cricket are cut, so that no sensory information can pass into or out of the cricket's body, its motor nerves will still continue to fire in the characteristic pattern of a calling song (Bentley and Hoy 1974).

The cricket song is one of the finest and most complete examples of a genetic pathway. It was traced, bit by bit, from the full behavior in a natural environment back to a single type of brain cell. It is rare in behavior genetics that such a full cycle from behavior to genes to behavior has been so fully illuminated, so it has some important things to say about the genetics of behavior.

Biological pathways from genes to behavior are extraordinarily complex. Very little is known yet of the way in which genes are decoded into nerve and muscle so as to control or even influence behavior. Scientists sift through thousands of tiny clues from human physiology and reach tentative conclusions with the greatest effort.

But if work on humans is partial and fragmented, research on fruit flies and crickets is more precise. It is here that basic questions in behavior are posed and answered. How do genes control behavior? What channels do they use? What is their relative influence on the environment? Insects are a long evolutionary step from humans, which is precisely why they are useful. Their simple physiology and primitive behavior make it possible to uncover genetic pathways that are far more intricate and inaccessible in more complex animals.

In the cricket, the genetic pathway leads from a single command nerve cell that originates in the cricket's tiny brain to the motor neurons that work the muscles, in a precisely programmed sequence to the muscles that drive the wings, to wing movement, to orchestration of the entire song. The only step that remains to make the pathway complete is to find out just how the genes have installed their program in that neuron. It is known that many genes, not just one, are behind the cricket song, but it is not clear how this polygenic system is channeled through one neuron. What does that neuron do when it reflects a genetic change from one species of cricket to another?

In their article in *Scientific American*, Bentley and Hoy suggest that the genetic code could take the form of changes in the firing rate of the command neuron. They found, for instance, that the only difference in song between two wild species was one trill. One species emitted a song with two trills, the other with three. The distinction could be traced to a single nerve impulse—an infinitesimally small increase in the firing rate of the neurons. That one impulse was enough to keep the species reproductively isolated from each other, and, as the authors point out, it is a "remarkable example of fine genetic control."

Critics of behavioral genetics like to mention that man is not a cricket and that such example of instinctual behavior are of little value in understanding human behavior. The normal person under normal circumstances rarely acts instinctively. Almost everything the human does is affected by learning and experience, as well as by genes.

But the analogy between crickets and man needn't be made at the level of behavior. Humans don't make music with wings either; they attract their mates in other ways. We can learn important lessons from crickets, fruit flies, mice, and gerbils nevertheless—not about behavior per se, but about the mysterious biological pathways that lead from genes to behavior.

INSTINCTS NOT ENOUGH

The cricket song stands at one extreme end of behavioral genetics, where stereotyped, innate behavior is locked into biology by natural selection to ensure the preservation of the species. Observers of animals once believed that this instinctive type of behavior controlled the responses of animals much more completely than more recent studies indicate. A closer look shows less and less instinctive behavior. Increasingly, animals above the level of insects show genetic channeling of behavior and also much room for the environment and for learning to have an effect.

This is particularly true of mammals, the highest evolutionary order of animals, which includes man. Once animal life crossed the mammalian threshold, it entered a new world of brain complexity which radically altered the nature-nurture balance. Although still genetically based, behavior became far more malleable, more subject to learning, less automatic.

The gerbil offers some insight into the nature-nurture balance of a mammal. It does much of its communicating by scent, and on its belly it carries an important scent gland, full of volatile sprays and odorous chemicals called pheromones. The pheromones secreted and spread by the gerbil accomplish sex signaling, trail marking, or dominance advertising.

Traditionally, pheromone communication has been considered completely innate behavior and, by the classical definition, a pheromone is, among other behavioral stimuli to individuals of the same species, a sex signal, programmed for automatic action on the part of both sender and receiver.

But, according to University of Texas scientist Dr. Delbert Thiessen, not all behavioral geneticists believe this any longer. Dr. Thiessen's work at Texas indicates that gerbil pheromones are not coded for specific behaviors. On the contrary, the pheromone system is a communication channel more like a very primitive vocabulary that can be used to signal several messages, depending on the learning of the animal. Laid down in one context, the pheromone might mark a trail; in another context, the same chemical may alarm or attract an animal. If no critical social event is tied to the signal, the gerbil will simply smell and then ignore it.

"The signal is potentially there to be used in many different ways," said Dr. Thiessen. "The foundations of the system—the ability to produce a signal, lay it down, and control its release through hormones—these things are coded in genetically. But aside from that, you have a host of possibilities for its use, depending on what the animal learns about that signal. Most mammalian communication systems are like that."

Although not all of the 50 to 100 separate compounds in the scent gland have been tested yet, it's possible that some could be coded for specific behaviors, not all of which have been tested. Also, each gerbil has a unique package; evidently slight alterations in the chemical balance signal individual differences and perhaps age. But even individual uniqueness is not immune to environmental change.

For instance, most recently, the Thiessen group has discovered that gerbil odors may reflect what they eat. Diet appears to make important changes in the pheromone—so important that pups actually turn away from adults which have been fed a diet different from that of their parents. Pups reared by parents fed Purina lab chow were given a choice of approaching two separate sets of unrelated adults—one of which had been fed the parents' diet, while the other had been given some different food. The pups approached the first set, but nearly ignored the second and split their preference roughly 75 to 25 on 100 trials (Skeen and Thiessen 1976).

Dr. Thiessen believes that every odorous quality the gerbil has is changed somewhat by the diet it consumes. In this case, not only do genes have no specified response but their primary action builds a flexible metabolic system that is altered in fundamental ways by its surrounding environment.

This kind of adaptiveness is useful to a species, it seems. A population which cannot adjust to environmental changes risks elimination in the evolutionary selection process.

MANY PATHWAYS, MANY SITES

It is becoming apparent from animal studies like these that genetic pathways to behavior may focus anywhere in the body, from brain cells to toenails. Whatever is physically based is genetically influenced; however, it is difficult to find the physical locus in the body which is having an effect on a given behavior. Behavioral genes may express themselves directly in the construction and maintenance of brain cells (e.g., cricket songs), or they may take more complex routes to less specific behavior, as through the gerbil's gut.

The Case of the Mutant Flies

Moreover, the actual site of genetic influence can be distant from where the behavioral or physical effect is manifested. This can be true of physical and mental diseases, as well as of normal behavior. Dr. Seymour Benzer, of the California Institute of Technology, who has done novel and important work with fruit flies, notes that, in certain cases of retinal deterioration in man, the problem is not in the eye at all but in the inability of the intestines to absorb vitamin A from food. Many cases of mental retardation can be traced to some metabolic defect which damages the nervous system by creating a chemical imbalance. PKU is the result of such a defect, and, while it is expressed through the brain, it is not manifested there primarily but, as we have seen, in a failure throughout the body to convert phenylalanine to tyrosine.

So difficult is it to find the locus of a genetic defect in behavior that Benzer resorted to extraordinary genetic techniques. His aim was to find the precise pathway leading to the strange behavior of mutant flies, insects with a single gene mutation that caused some obvious behavioral quirk (Benzer 1973).

The fruit fly, *Drosophila melanogaster*, has been subjected to years of genetic research, and its species is populated with dozens of mutant strains, some with white eyes, others with forked bristles or yellow bodies, some that can't fly, others that can't climb. The names of the strains reflect their defects: there are *Sluggish, Hyperkinetic, Drop Dead* (it lives for a few days and then turns over and dies), *Flightless, Wings Up* (after birth, its wings go straight up and stay there), *Comatose, Stuck* (inability to disengage after copulation), and *Coitus Interruptus*, among others.

Because the insects had been mutated by radiation or certain chemicals and then inbred to reveal behavioral defects, geneticists knew that genes were causing these defects. But they did not know how and where the genes were acting until the Cal Tech work.

Benzer first linked up the particular behavioral defect with some physical marker, such as white eyes and yellow bodies, so that both were on the same sex (X) chromosome and could be transmitted simultaneously into new progeny. Next, with a manipulation of the sex chromosome, Benzer created mosaic flies, with bodies half normal and half abnormal, reflecting not one genotype but two. The aim was to use abnormal color as a marker for the behavioral trait. The flies showed an unusual combination of yellow patches. Some had normal heads on mutant bodies; others were mutants on only one side, while still others had mosaic eyes, with white cells mixed into the normal red color.

With these strange insects, Benzer then used the mutated patches to lead him to the physical source of a behavioral trait. *Wings Up*, for instance, might be caused by mutations in the nervous system, in the muscles that control the wings, or in the wings themselves. Study of the mosaics led Benzer to the *Wings Up* fly's thorax because that section was more often made up of mutant tissue and it seemed to be a probable locus for the defect. Sure enough, dissection showed some of the flight muscles in the thorax to be highly abnormal. Fibers had either degenerated or were missing altogether, so that the fly could not pull down its wings. Thus, the *Wings Up* trait was traced to a genetic mutation, which produced the physical abnormality, which in turn produced the trait.

The *Drop Dead* mutation was quickly traced to mutated heads, indicating brain involvement. Dissection revealed striking deterioration. The brain was full of holes.

The *Hyperkinetic* mutation posed more difficult problems. This fly is so hyperactive that its legs shake even under anesthesia. Mosaic flies shook some of their legs, but not all, and in many cases the shaking was associated with a mutated body surface. Surface features such as color do not reflect interior parts, and clearly the defect was located somewhere inside the fly's body. But where? There were no markers for internal features. Here Benzer resorted to an ingenious solution which allowed him to trace the abnormal locus to a given region of the blastoderm—the embryonic cells of a new fruit fly growing in the yolk. Based on maps of the blastoderm, Benzer was able to conclude that the *Hyperkinetic* mutation was affecting an embryonic region known to be the source of the ventral nervous system. The fly had bad nerves.

The Cal Tech work dramatically and visually illustrates the power of genes in behavior. If a fly drops dead without apparent cause, or doesn't perform the courtship ritual, or lacks the circadian rhythm, one look at its patched body provides a visible clue: for example, it has a yellow head.

Every cell of the fly's body carries the defective chromosome with two mutated genes—one for the surface feature, one for the behavioral defect. Not every cell, of course, uses both genes, and not every Yellow Head manifests the defects. The behavioral defect expresses itself only at relevant locations, as in the nervous system. Still, the total genome—the complete array of genes and chromosomes—is packed away in each cell of an animal's body. What is known about that genome? How many genes does it contain? Can anything be said of its function?

A virus has the simplest genome on earth. It is, in fact, little more than a set of genes enclosed in a membrane which act by infecting a living cell and taking over some of its genetic instructions. Without a living cell through which to express itself, a virus appears to be dead. By itself, it stands on the threshold of life—halfway between living and nonliving matter. The smallest virus has five genes, and it is no accident that this primitive form of life is a bit of genetic material.

HUMANS AND DNA

At the opposite evolutionary extreme stands the human species with enough genetic substance in every cell for about five million genes. If this DNA (deoxyribonucleic acid) were stretched to its full length, so that the biochemical instructions followed each other in single file, it would make a chemical strand 2 meters long (Bodmer and Cavalli-Sforza 1976, p. 132). But in the cell nucleus, the DNA is coiled and folded back upon itself time and time again to build a dense three-dimensional structure of incredible complexity.

Man does not actually have five million genes, although he has the DNA for it. As much as 40 or 50 percent of human DNA is repetitious; that is, the same genetic chemical sequences appear over and over again. The function of the extra material is unclear. The rest of the DNA does contain unique sequences—ostensibly the instructions for different genes, but how many genes? Estimates vary widely. At the outside, the human could have 2.5 million genes, but most geneticists favor a lower figure, the lowest being 100,000, a figure that reflects the number of enzymes and other products of genetic instruction that are thought to be involved in human physiology.

The Alleles

A great many of these genes exist in several forms. There may be 2, 3, 6—even 10 or more—different forms of one gene, called "alleles," that can occupy the same genetic location. Thus, it isn't strictly accurate to speak of genes, but better to speak of gene loci or gene functions. The actual chemical sequence which directs the production of a given enzyme may occur in several allelic forms, each of which does the work a little differently.

The existence of so many alleles is a source of great genetic variety in man as well as in other animals; the possibilities for individual and racial variation are great indeed. In some instances, the gene variants may reflect geographical location, which indicates that some genetic adaptation to local conditions has taken place. But more often there is no explanation. A major issue in contemporary anthropology is the meaning and purpose of all this genetic variation.

The existence of so many alleles at one locus may simply mean that pressures of natural selection have not been severe enough to keep a particular gene sequence intact, so that many variations occurred, none with special survival significance. It is believed that some 30 percent of the human genome has two or more alleles at each gene locus; other mammals have shown similar amounts of variation. What impact such variation might have on behavior is a matter of speculation, but two points can be made:

(1) Traits that are crucial to survival tend to have little or no genetic variation because the forces of evolution have fixed specific gene forms. Any deviation from the crucial sequence usually invites disease or death.

(2) Behavioral studies typically show large degrees of genetic variability (Thiessen 1973). Thus, it is possible to believe that natural selection has not been acting forcefully on much of behavior. Perhaps the genetic variability that underlies individual differences has little or no evolutionary significance for the species except at the extremes of a given behavior. It may, nevertheless, have a great deal to do with the individual's personal comfort and adaptation in a given environment.

Allelic differences are not the only source of genetic variation. They may not even be the most important source. In the early sixties, two French scientists, François Jacob and Jacques Monod, introduced a new genetic concept that would lift behavioral genetics onto a new plane of understanding. Before their discovery, the genetic code looked relatively static and its role in behavior seemed more inflexible. Genes dictated that certain proteins be made and that those proteins then act upon body tissue and brain. Genetic action moved in one direction and, despite its great variability, the code was still a dictator protected from change or influence by the creature that carried it.

The original discovery by Jacob and Monod was made in *E. coli,* a common bacterial inhabitant of the human intestine. One of the genes in *E. coli* is responsible for producing a sugar enzyme called "galactosidase." But the action of that gene can be suppressed or raised a thousandfold by a second gene, the regulator, which is sensitive to the amount of certain sugars already in the cell. The discovery indicates, in bacteria at least, that by this means, the genetic code recognizes and adapts to changes in its environment. Many researchers feel that more evidence is needed in mammals (Orten and Neuhaus 1975, p. 119). Since regulator genes appear to be responsive to the environment, they can be affected by the state of the organism at any given time, by its cellular environment, and perhaps by such physiological states as circulating hormones, although there is no direct evidence to that effect. Thus, the genetic impact is brought under a whole set of influences. It becomes dynamic and responsive.

The Regulators

It now appears that there are two major types of genes with quite different functions. One type, the structural, directs the production of specific protein, including the thousands of enzymes that run body processes. Another type of gene regulates the first. It produces no protein; instead its role is to tell the structural gene when to act and for how long. A regulator gene may control just one structural gene or an entire bank of genes; it may call them into play depending on the needs of the organism.

Much of *human* development, in both physical growth and mental processes, is more comprehensible in the light of this theory of regulator genes because it illuminates how genes and environment may be working together to produce a unified effect. When growth is retarded because of malnutrition, one can visualize the regulator genes inhibiting the production of bone-building protein in response to the nutritional state of the cell. When considering personality and intelligence, one can imagine regulator genes controlling the levels of hormones and brain chemicals critical to emotional and intellectual functioning and moderating those levels according to individual experience, such as harsh rearing, soft rearing, amount of stimulation, and a host of other living experiences. It is to the provisional illumination of these far more complex and elusive pathways in humans that we now turn.

Chapter IV. Potential Pathways in Man

It has been said that the only instinct left to the human race is the startle response. All else is affected by learning and setting.

It is true, indeed, that the human brain is not designed to produce automatic, instinctive behavior exclusively. A reptile's brain does that. It responds to the environment out of innately programmed nerve cells located in the ancient brain core.

According to the well-known thesis of NIMH scientist Dr. Paul MacLean, that reptilian brain still resides in man, but it is covered over and constantly moderated by two additional layers of neural organization. The human brain, then, has a three-tiered organization, and nerve signals moving in or out of the nervous system pass through all three tiers.

Above the reptilian brain is the old mammalian brain, which Dr. MacLean has dubbed "the limbic system," a primarily emotional brain which he speculates arrived with the appearance of the first mammals. Atop that is the cerebral cortex or new mammalian brain which ballooned to its present size during human evolution. Thus, in man, the instinctive reptilian brain has no direct route to behavior. Its messages must go the long route through higher centers.

But diminished control by the instincts does not mean freedom from all genetic control. Genes operate through many pathways in the brain, some of which will be described here. They direct the production of protein from which the brain is constructed and by means of which memories are imprinted. They play a role in controlling the level of important biochemical transmitters, and they produce a range of interesting substances of a class called "peptides" which seem to affect behavior and which may become incorporated into hormones.

THE PEPTIDES AND THEIR ROLES

A peptide is a fragment of a protein molecule, made up of a string of amino acids, like pearls on a tangled necklace. One or more genes can produce a peptide, which consists of chains of as many as a hundred amino acids, still far smaller than most protein molecules. Until recently, peptides were considered merely the building blocks of protein; now it seems that some of them may have behavioral relevance by themselves. Several peptides have been found that may influence feelings and behavior. Among these, morphine-like peptides called endorphins appear to be involved with analgesic effects and may play a role in narcotic euphoria and dependence. They bind with specific receptors in many parts of the brain and elsewhere, not at just one center, and are produced naturally by the brain. Two of these, only 5 amino acids long, are called enkephalins; others are much larger. Thus, humans normally produce morphine-like substances, a string of amino acids, whose precise function is still uncertain but which may have an important effect on emotion and feeling states.

Another kind of peptide with effects on behavior is a fragment of a very important hormone secreted by the pituitary to activate the adrenal glands when the body is responding to fear or challenge of some kind. The hormone, ACTH (adrenocorticotropic hormone) is a molecule of 39 amino acids, but the first 4 to 10, by themselves, are enough to bring about important effects on animal and human behavior.

A main function of this fragment appears to be to assist memories, perhaps especially of an unpleasant event. Animals given this peptide do not forget, at least not quickly, what they have been through in the past. This observation is graphically illustrated by an account of work done by David DeWied, a Dutch pharmacologist at the University of Utrecht in Holland, where rats taught to jump from a pole to avoid electric shock remembered the lesson far longer with an injection of ACTH than without it. And long after noninjected animals had forgotten the association, ACTH-hyped animals still jumped from poles in anticipation of shock.

In human volunteers, the ACTH peptide seems to improve learning or memory. With the peptide booster, people seemed less anxious while taking a test, at the same time that their visual memory was improved. In another study, ACTH improved reaction time. This could mean that ACTH, particularly the set of the first 10 amino acids of the hormone, is a biochemical mechanism for arousal and that its effect on learning is simply that people remember better when aroused than they do in a less motivated state. Arousal and motivation are recognized as crucial to efficient use of the intellect.

Dr. Seymour Levine of Stanford favors the arousal theory. Taking a conservative position, not attributing too much or too specific a function to ACTH, Levine believes the ACTH peptide has clear behavioral effects, but he thinks its effects are generalized to provide arousal in a wide variety of situations.

Other scientists think the peptides might be coded for more specific behaviors. Dr. Delbert Thiessen suggests that the peptide ACTH might be coded for fear—that it provokes fear, which is then matched by adrenal outpourings which energize the body to act and to counteract the emotion (1972).

This theory is based on evidence that rats with their adrenals removed become more fearful, while rats with their brain source of ACTH removed become less fearful, in the same threatening situation (Weiss et al. 1969). The endorphins and the ACTH fragment are only two examples of peptides in the brain and pituitary which are turning out to have new and unsuspected functions in mediating behavior. Moreover, they appear to share a common precursor, for which the name "prooprocortin" has been proposed.

Thus, these peptides alter behavior in ways still only dimly seen. And there may be other peptides not yet isolated with functions still unspecified. Moreover, it could be that the peptides integrate themselves into more than one protein molecule, so that the same behavioral element may be evoked by different proteins. On the basis of ACTH results, Thiessen suggests that scientists look for other amino acid chains with different effects on mental processes, with the hope someday of detailing an "amino acid map" for behavior (1970). But dramatic as they seem, the peptide results raise difficult problems of interpretation—problems that plague everyone working at the frontiers of the biology of behavior. What is behavior biologically? How is it represented in the brain? Where is it housed? How specific are these peptides in terms of the behavior they influence? Influencing specific functions would imply behavior is more thoroughly coded into DNA than anyone has yet realized.

THE NEURONS AND WHAT THEY DO

If modern neurophysiology is any guide, behavior is not carried by chemicals, either peptides or hormones; it is more probably housed in the anatomy and organization of the brain, and while neurochemicals may modulate behavior, they don't produce it.

A century of surgical work has suggested that both memory and behavior are physically housed somehow in brain neurons and their organization. Many a patient, while undergoing brain surgery on the operation table, has relived past experiences when specific brain areas were stimulated. The brain has no pain endings and surgery often requires that patients be conscious so that their neurological functioning can be continuously checked. In that state, people have heard music, seen images, experienced emotions—all due to an electrical stimulus in the right set of neurons.

Many sorts of behaviors and attitudes—sleeping, eating, fighting, thinking, meditating, hoping, fearing, feeling good, feeling bad, even experiencing orgasm—can be produced with an electrode in different brain regions. The well-known brain scientist, José Delgado, dramatically illustrated this fact with his famous experiment on a raging bull. With nothing more than a remote-controlled electrode implanted in the animal's brain, Delgado brought to a stop a bull that was in full-scale attack. But Delgado also demonstrated finer control over behavior with some less well-publicized work on monkeys. Implanting several electrodes in specific sets of brain cells, Delgado successively stimulated each one and produced disconnected bits and pieces of behavior—a grin here, a raised eyebrow there, a flicker of the eyelid, a twitch of the nose. One by one, the components of an emotional response appeared on the monkey's face, until finally Delgado stimulated the last electrode in a new area, at which point the monkey suddenly integrated all this behavior into a single purposeful act. The point seems to be that behavior is coded by bits and pieces in neurons and integrated through many pathways into meaningful action.

The construction of this substratum of behavior—the anatomy of the brain—is directed or "programmed" by genes. Genes specify the growth of neurons in the fetus so that, by birth, all neurons are basically intact. They direct that the human brain be organized in certain ways and that it have certain important structures. At the base of the brain is the brain stem, containing structures that control autonomic body functions such as heart rate and respiration. Above the stem is the limbic system, considered the seat of emotions, where love and rage can be experienced, where sexual, social, and hostile behavior can be produced. Limbic structures are not the only places that produce emotional behavior, but they are the most important. Finally covering all is the cerebral cortex—the so-called higher center of intellectual and sensory processes.

NEUROTRANSMITTER THEORIES

It is in this three-tiered anatomy with its major interconnecting pathways that genes have laid down their contribution to the basic biology of behavior. In addition to this anatomy, however, genes have a second and very important role to play in behavior. They help determine the concentrations of neurochemicals that wax and wane at different sites in these structures, in effect, turning neurons on and off, and thus affecting the final behavioral output. These neurochemicals have been the focus of biological research in mental illness for two decades.

The effects of the relative excess or depletion of these neurotransmitters underlie many modern theories of mental illness. Years of transmitter research have suggested that an important role is played by neurotransmitter amines in depression, mania, and schizophrenia and in other less common forms of mental disorder. No theory of brain function has been more productive than transmitter research. It forms the basis of much drug treatment of behavioral conditions. In turn, exploration of this pharmacologic action provides the basis for many of the biochemical hypotheses about serious mental illness.

Transmitters act at nerve synapses to transmit activity from one nerve cell to another, or, alternatively, to block the transmission. An activated neuron may release a transmitter from its nerve endings, which then turn neighboring neurons on or off, depending on the type of transmitter released. This alternating activation and inhibition of nerve cells is a primary principle upon which the brain modulates its functioning.

Transmitter levels may be modified by drugs with pronounced effects on behavior. The physical anatomy of the brain can be altered only by physically destroying brain cells; but the neurochemistry is rapidly and temporarily altered by a variety of methods using interventions as different as drugs, nutrition, physical activity, stress, and electric shock therapy. Many of these transmitters are already known and more are being discovered. Whether general or more specialized in function, they are affected differently by various pharmacological agents. Thus, the potential for improving brain function and treating mental disorder through biochemistry is great.

A few examples illustrate the point:

Serotonin is a transmitter closely involved with sleep and sedation. Found primarily in the brain stem in a nerve complex called the *raphé nuclei,* serotonin travels upward through pathways that fan out to the hypothalamus and parts of the limbic system and cerebral cortex. At sites throughout the brain, serotonin inhibits nerve activity. Put simply, serotonin seems to shut down waking activities. For instance, a cat deprived of the source of serotonin does not sleep well. Ostensibly, serotonin, or rather the lack of it, could be a cause of insomnia. Unfortunately, efforts to raise serotonin levels experimentally and thereby combat insomnia have not yet been very successful. One of the methods scientists have tried for improving sleep in humans is the use of the precursor amino acid tryptophan (which is abundant in milk and peanuts, and which the body changes into serotonin) but results are subtle and equivocal. Most recently, serotonin has been tied to depression, and it does appear from work at NIMH that some depressed patients are serotonin-deficient. Increasingly, then, this transmitter is seen as having a role in maintaining mood level.

Norepinephrine (NE) looks at first blush like an opposite to serotonin—at least in terms of its role in behavior. For years this transmitter has been linked with arousal, aggression, physical action, and heightened mood. It has been considered an important key to depression, i.e., low levels of norepinephrine in the brain coincide with a depressed and apathetic condition. Supposedly, the major antidepressant drugs used in treating humans act mainly by raising norepinephrine levels. So does electric shock therapy. But lately the story of norepinephrine has been getting more complicated. Further experiments indicate that it does not seem to be an antidepressant per se: the chemical cause of arousal. Rather it seems, as do all transmitters, to have its effects on behavior because of the area of the brain it acts upon. Like serotonin, norepinephrine may be mainly an inhibitor of neural activity, and one speculative explanation for its effects goes like this: Norepinephrine shuts down another chemical system which inhibits limbic structures. Thus, its total effect is to arouse or disinhibit behavior, apparently by suppressing the normal controls on behavior that operate in specific regions of the brain.

Another explanation, quite the opposite, suggests that norepinephrine could act by keeping limbic structures from going haywire from overexcitability. Too much activity could, in effect, blow the system, leading to a shutdown of neural communication and subsequent severe catatonic depression. Raising norepinephrine levels would then reinstitute normal modulating controls. Actually, both explanations and several others are possible, depending on the behavior in question and the region of the brain one chooses to study.

An animal study which bolsters this second explanation is the audiogenic seizure (a seizure triggered by a loud noise) in rodents. Audiogenic seizures are associated with very low levels of norepinephrine in rats, mice, and rabbits. Moreover, the rodents can be protected from seizures by raising norepinephrine levels (Schlesinger 1970).

Nothing seems more akin to blowing the system than an audiogenic seizure. The susceptible mouse, placed in a bell jar, hears a loud bell ring and for a moment nothing happens. Then it begins a wild running around the jar, in stiff-legged bounds. In the next phase, the mouse falls on its back and goes into myoclonic seizure, while its legs pull rigidly into its body. Then they rigidly stretch out in the opposite direction in a myotonic convulsion. At that point, the seizure invades the animal's respiratory center and it dies from paralysis of the nerve cells, unless something is done to keep the mouse breathing until its nerve cells recuperate. Schlesinger preserves the lives of his experimental animals with what he likes to call "mouse-to-mouse" resuscitation.

Schlesinger believes norepinephrine protects the nervous system, particularly the limbic structure, from over excitation, but he said this cannot be the only explanation for audiogenic seizures. Some other genetic susceptibility besides low norepinephrine levels is involved in causing this disorder.

Dopamine is currently one of the favorite scientific villains for explaining schizophrenia.

Though serotonin, norepinephrine, and dopamine are less abundant than other neurotransmitters, dopamine is more common than norepinephrine, which is a further chemical derivative of dopamine. Its main pathway leads from the *substantia nigra* in the midbrain primarily to the *corpus striatum* and out to other centers. One of these pathways enters motor areas and the availability of this transmitter has important effects on movement. Loss of dopamine causes Parkinson's disease, apparently because in the absence of this transmitter, which inhibits nerve cell activity, movement can no longer be modulated. The results are the tremors and rigid muscles of Parkinsonism. In addition to these problems, Parkinson's patients often show mood disturbance and "flat" facial expressions, indicating a second pathway for dopamine—directly into limbic structures crucial to emotional behavior. As a third route, dopamine affects activity in the hypothalamus, the brain center for controlling hormone release through the pituitary gland. Thus, dopamine may also play a role in hormone regulation.

Current theory suggests that schizophrenics may have an excess of dopamine. Where and how this excess might be acting to produce the bizarre symptoms of schizophrenia is unknown. The theory, however, is based on the fact that drugs useful in treating schizophrenia are effective mainly because they block dopamine activity. But the side effect of these drugs is sometimes to produce Parkinson's symptoms, and scientists have not yet found any way to get the benefits of these antipsychotic drugs without also producing a movement disorder in some mental patients. The trick is to find a way of blocking dopamine activity in some regions of the brain without blocking it in motor areas.

AMPHETAMINES AND THEIR ACTION

In an opposite effect, amphetamine drugs offer a mirror image of the dopamine process. Rather than curtailing dopamine activity, they enhance its flow, with the result that amphetamines can relieve some movement disability; they also can make some people psychotic.

Amphetamines are among the most intriguing and paradoxical drugs in modern research. They bring about an interesting assortment of behaviors—stimulation, arousal, enormous physical energy far past the point of normal fatigue, suppressed appetite, and even enhanced learning. Finally, as with all psychomotor stimulants which release neurotransmitters, amphetamines can produce a condition difficult to distinguish from paranoid schizophrenia, which has been labeled "amphetamine psychosis."

In the 1930s, amphetamine was found useful in the treatment of narcolepsy, and it has often been prescribed for obesity. Throughout the years, as cases of amphetamine psychosis were reported, clinical studies were conducted showing that the drug could indeed produce paranoid psychosis, with ideas of reference, delusions of persecution, and hallucinations both auditory and visual. Many cases were resolved within a few days of drug withdrawal, while others continued longer with some residue of disorder. Still other patients who indicated a history of amphetamine use continued in a paranoid schizophrenic state for months or even years, presumably a schizophrenic episode precipitated by the drug (Davis 1975, pp. 136-137).

Somehow, excess dopamine seems to be involved in all of this. How and why this is so are among the most pressing questions in biological psychiatric research. Does it inhibit normal controls on emotional behavior? Does it shut down sensory gates so that people are no longer so aware of what goes on around them and incoming messages become distorted? Does it cause people to be underaroused or overaroused by the environment? An answer to any of these questions would constitute a major advance in mental health.

Recent researchers are postulating amphetamine involvement with the other transmitters as well. For instance, they point out that amphetamine enhances the effect of central reward mechanisms—the amphetamine addict is an obvious example of the rewarding action in man. Thus, the facilitatory effects of amphetamine on behavior may be mediated through the release of the neurotransmitters: in the case of reward and of addiction, by norepinephrine; in the case of punishment, which can include the suspicion and fears of paranoia, by serotonin; and, in the case of stereotyped behavior, by dopamine in high dosages (Stein and Wise 1973). Others agree that consideration should be given to the evidence that there is inhibitory involvement of serotonergic fibers in the behavioral actions of amphetamine. They see this as suggesting that these fibers are not only involved in the symptoms of amphetamine psychosis but may also help to explain the inhibitory actions of amphetamine in the treatment of hyperkinetic children (Breese, Cooper, and Hollister 1975, p. 130).

Contradictions, Revisions, and Advances

As neurotransmitter work continues, scientists are going far beyond the first crude assessment of neurochemistry and behavior, and the mechanisms of action are found to be more complicated. Contradictory findings in many of the results mean that transmitters must be reevaluated. Norepinephrine is no longer just the chemical of arousal; dopamine is no longer limited to Parkinson's disease. Researchers are still seeking to learn more about the involvement of the neurotransmitters in synaptic activity and how they interrelate with each other. They are finding it necessary to revise the early simplified models, which seemed to indicate that a specific neurotransmitter controlled a specific behavioral activity.

Scientists are finding a revised and even broader role for transmitters in behavior. The fact is that quite small changes in the level of these amines do influence behavior in ways that would seem to affect an individual's ability to adjust to his environment. The effect is probably important for normal as well as abnormal behavior. One of the enthusiastic pioneers in the field, Dr. Daniel X. Freedman, is among those taking a new look at the problem.

Illustrative Animal Studies

Dr. Freedman's studies are showing that minor depletions in the amines will cause animals to become abnormally sensitive to a drug like LSD and probably more sensitized in general to all stimuli. No more than a 15-percent depletion of norepinephrine and serotonin was needed to make Dr. Freedman's rats more than twice as sensitive to LSD as they otherwise would be. A 15-percent depletion is well within the average range of variation for levels of these neurotransmitters in rats and probably humans as well.

Usually a rat will not respond to LSD in doses of less than 100 micrograms per kilogram. But the supersensitized rats with minor amine depletion responded to only 40 micrograms. And long after the animals looked recovered—as long as 2 weeks after the initial depletion, when amine levels had recovered to about 85 percent of normal—they still were supersensitive.

Dr. Freedman said this same enhanced sensitivity to drugs can be found in humans with low transmitter levels.

"So what does this mean?" he asked. "It means we can expect the amines to play some sort of filtering or buffering role and maybe the deficient animals are at some sort of disadvantage. It's interesting that you don't have to be much depleted to show some rather startling effects."

Until recently, there wasn't much genetic information on the brain amines. People assumed that the system is affected by genes; all biology is. Moreover, in the mid-sixties there was evidence that brain levels of these amines varied widely in different strains of mice and rats. But the evidence that genes function importantly in controlling amine levels has only recently been shown at Stanford University by a team of scientists including Drs. Jack Barchas, Roland Ciaranello, and Seymour Kessler. Their approach was to study levels of enzymes known to be involved in the synthesis of transmitters. Low enzyme levels would indicate low amine levels because the transmitters cannot be generated in the absence of these critical gene-produced catalysts, the enzymes.

"I think we have established that there are very strong genetic controls on these systems in terms of the amount of enzyme available in the brain to synthesize amines," said Dr. Barchas.

The only way really to prove genetic control is to crossbreed mice with different enzyme levels and then measure the levels in the offspring. The pattern of resemblance indicates not only that genetic control is present, but whether the inheritance is passed on by dominant genes, co-equal genes, or many genes.

If dominant (and recessive) genes are involved, then the offspring of two different lines will resemble one parent, as when brown- and blue-eyed parents produce brown-eyed children. If co-dominant or co-equal genes (no recessives) are involved, the offspring will fall in the middle between the two parents, as when red and white sweet peas produce pink sweet peas. When many genes are involved, distribution of offspring of the second filial generation will range from one parental extreme to another (and even beyond), as when parents of different heights produce progeny of widely varying heights. Typically, behavioral traits turn out to be polygenic systems with many genes involved, which makes genetic analysis difficult.

In the case of the brain amines, crossbreeding produced some intriguing results. Two mouse strains, when bred to each other, produced intermediate values in the offspring, suggesting polygenic inheritance or no dominance. But then a third strain was bred which started with intermediate values. This time, offspring resembled one parent—the one with middle values, while the genes of the parent of either extreme were not expressed. Stanford researchers believe the third strain carried dominant genes which provided moderate levels of enzymes critical to amine synthesis (Kessler et al. 1972).

In addition to the transmitters, Stanford scientists found that genes were regulating still another very important chemical system in the brain. This chemical, called "cyclic AMP" (cyclic adenylic acid), is a common final pathway for expression of the transmitters. It directly inhibits the firing of nerve cells, and when serotonin, norepinephrine, or dopamine cross the synaptic gap, they affect production of cyclic AMP.

Since cyclic AMP production at the synapse is a final step of the pathway of nerve cell activity and ostensibly is affected by many influences, both genetic and environmental, it surprised the Stanford scientists to find a clear genetic influence operating at that point. But there it was. Four strains of mice—two with high cyclic AMP, two with low levels—were interbred and the offspring followed the low-level parents. Again there is a suggestion of dominant genes acting, this time to maintain cyclic AMP at the lower values (Barchas et al. 1974*b*).

For the most part, this genetic work on the amines has not yet been associated with behavior. There is some suggestion that mouse strains with high norepinephrine are more aggressive than others, but the associations are not terribly high. The field of behavioral genetics may be due for a major advance in this respect.

The Pathway for Aggression: Cases in Point

Three years ago, two independent laboratories—one at the National Institutes of Health, the other at Stanford—found a form of aggression in mice that was clearly inherited as a single recessive gene. Nothing like it had been found before: a complex behavioral trait controlled by one gene at that. The proof is solid because, in crossbreeding experiments, offspring turned up with a classic, single-gene Mendelian distribution.

The distribution looked like this: When two lines were crossed—one aggressive, one docile—the first generation of hybrids was docile. The docile trait was dominant. In the second generation, the genes reassorted themselves in classic fashion, producing one aggressive mouse (recessive genes teaming up) and three docile animals (two sets of mixed genes and one set of purely dominant genes). Dr. Roland Ciaranello, who was at NIH at the time, was stunned by his results. He tested 80 animals before he could accept the evidence. Then he called a colleague in California.

"It was a mindblower. I called Kessler at Stanford and I said, 'Seymour, you're not going to believe this, but I have this crazy result and I don't know what to do with it.' He started to laugh because he had been doing the same study with different mouse strains and he was getting exactly the same results."

The aggressive mice showed one clear difference from the docile mice: They had higher levels of cyclic AMP. Somehow there was a link between the recessive trait and high cyclic AMP, but how direct or indirect that link might be is still unknown. The pattern of inheritance for cyclic AMP does not resemble the profile for this recessive gene; thus, one cannot be considered the cause of the other. Nor is there yet a well-founded theory to explain why high cyclic AMP levels should be associated with aggressivity, although it is consistent with the transmitter story.

Both Drs. Ciaranello and Kessler hasten to add that they found this pattern for only one form of aggressivity in mice, not for aggression itself, which takes many forms. This type of aggression is provoked by keeping the mice in isolation for 2 weeks and then placing a strange male in the cage. Isolation is profoundly stressing and many animals will immediately attack a stranger when they are finally given company. The two breeds tested here were closely related, but highly dissimilar in this one behavior. The aggressive-strain mice would attack usually within 10 seconds after seeing the strange male. The docile-strain mice would not attack at all, no matter what was done to them.

One might think from this experiment that isolation was more at issue than aggressivity, except for one fact. This particular strain, the BALB/cj,* is infamous for its fighting behavior; in fact, the laboratory that raises these mice usually warns its scientific customers about their behavior. The males are so aggressive that they will attack and kill each other even without isolation, unless they have been raised from birth in the same cage.

Now the objective will be to work out the genetic pathway leading from gene to behavior, and that won't be easy. Scientists now have a behavior that is clearly genetic and a neurochemical index, the level of cyclic AMP production, that is clearly influenced by genes, but, to pin the two together, they must find a region in the brain where the action is occurring. Overall the levels of chemicals in the whole brain do not tell very much about behavior.

"If we can find the right area to look at, I think we will find a chemical balance there which is determining how aggressivity is expressed. But to find that . . . where do you begin to look? I'm open to suggestions," said Dr. Ciaranello.

Other scientists have been trying to replicate this potentially rich experiment. An illustration of the tribulations researchers face can be drawn from the caution expressed by Dr. Donald

---

*Stands for B(Bagg) ALB(albino) c(subline) j(Jackson Laboratory).

Reis, of the Laboratory of Neurobiology at Cornell University Medical College's Department of Neurology. He mentions the difficulty of establishing "proof" of the presence or absence of a specific neurotransmitter as the cause of a specific behavior or disease. Many criteria, involving many techniques and disciplines, must be met, and account must be taken of the "fact that behavior is organized in *networks* of neurons deployed in interlocking chains." Another difficulty is that of defining precisely the behavior or disease being studied.

His case in point is the attempt to analyze the relationship of specific neurotransmitters in aggression, which has been defined as "behavior leading to the destruction of a goal object." Aggression, when it is looked at in terms of its *expression*, seems to exist in only two forms, affective (irritable) and predatory, each of which may be mediated by separate central neural pathways. The data on which he bases his behavioral hypothesis suggest that each type of aggression has its own distinctive neurochemical substrate. In short, the evidence for the hypothesis suggests that norepinephrine, and possibly dopamine, facilitates affective and inhibits predatory aggression; that acetylcholine facilitates both, particularly predatory aggression; and that serotonin, on the basis of fragmentary evidence, seems to suppress the expression of both modes of aggression (Reis 1975, pp. 146–147).

OTHER ENZYMES, OTHER ROLES

All the chemistry so far described concerns enzymes that promote the synthesis of brain amines. They help determine how much transmitter will be available in any given region of the brain because without these enzymes no chemical synthesis can take place.

But there is an entirely different class of enzymes which helps control amine levels and which also shows genetic influence. MAO (monoamine oxidase) functions as one of the main disposal mechanisms at the neutral synapse for serotonin and the catecholamines in both brain and body. It destroys these neurotransmitters by converting them into breakdown products which are then flushed from the body. MAO also destroys abnormal metabolites and functions in many other tissues besides the brain, although in different forms. If MAO activity is blocked, neurochemicals accumulate. In fact, blocking MAO activity is a good way to treat depression. An MAO inhibitor, discovered by accident years ago, was the first antidepressant, and it remains a common treatment for depression.

High levels of this degrading enzyme, MAO, mean that transmitters are being destroyed at a fast clip and that, depending on the rate of synthesis, lower levels of these neurotransmitter amines will result. Low amine levels, it should be remembered, are now associated with depression, and perhaps nervousness, among other traits of sensitivity.

By contrast, low levels of platelet MAO activity are thought to be associated with chronic schizophrenia, plus a range of stimulated and flagrant behaviors such as over-elation, mania, restlessness, insomnia, and paranoia. A low MAO level suggests that these transmitters are building up in the brain because the garbage disposal is not working at peak efficiency. High levels of dopamine are, of course, thought to be involved in the biology of schizophrenia.

Much of this behavioral evidence on MAO stems from 20 years of experience with a drug that lowers MAO activity. Iproniazid was first introduced in 1951 as a treatment for tuberculosis. When it was discovered that TB patients thus treated were becoming elated, the first antidepressant drug was born. MAO inhibitors like iproniazid were used for years in treating depression. They still are in use, but they do carry the risk of shooting a depressed patient to the opposite extreme.

On the face of it, pieces of the MAO puzzle appear roughly to hang together; high MAO/low-transmitter levels=depressed, over-reactive nervous system; low MAO/high-transmitter levels=stimulated, possibly crazy behavior. It must be kept in mind, though, that neurotransmission can be modulated by many factors, such as rate of synthesis, rate of degradation, and receptor sensitivity.

It is quite possible that MAO will indeed prove to be a "biological marker," as many scientists believe. Most of the work so far has been done with the rigorous testing of platelet MAO, which is formed from the solid portions of venous blood that has been mixed with a crystalline preparation and centrifuged. The MAO in human platelets has some characteristics in common with those found in other tissues, including the brain, and platelets are easily available for the study of monoamine transport, storage, metabolism, and receptor functions.

The findings so far have been interesting and fruitful. There is a low level of platelet MAO activity among 10 percent of normal people and a tendency for it to be reduced in the affective disorders (mania or depression) but not in unipolar depression (recurrent depressive episodes without mania). Forty percent of chronic schizophrenics show a reduced level usually more severely than in the affective disorders, but acute schizophrenics do not evidence this reduction. These differences are considered in more depth in chapter V.

Another interesting finding is a sex difference. Females tend to have platelet MAO activity about 10 to 20 percent higher than that for males. Many older persons show this tendency as well. The consequence of these differences has yet to be spelled out, but some scientists theorize that the higher MAO levels and hence lower transmitter levels in the aging and in women may account for the higher incidence of depression typically found among them (Murphy, Belmaker, and Wyatt 1974).

NIMH scientists Drs. Dennis Murphy and Monte Buchsbaum, working with normal college students, have derived recent data that add new weight to the MAO story. It looks as though MAO does indeed reflect some personality traits, including vulnerability to psychopathology.

The findings are in a sense extraordinary. That something in a bit of blood in healthy young people could indicate anything at all about personality is stunning. Yet there it was. On two common personality tests given to nearly 100 volunteers, low MAO activity, using both platelet and plasma amine oxidase, showed a significant correlation with relatively high scores in psychopathology. The findings were especially clear with a group of men who had very low enzyme function (Murphy et al. 1977).

With that, the NIMH team decided to sample blood from 375 college students, and, on the basis only of MAO activity, predict the risk for psychiatric difficulty. Results were positive. Individuals with low MAO activity reported more psychiatric counseling and more frequent trouble with the law than did students with high values. Moreover, the suicide rate among men with low MAO and among their family members was much higher— eight times as high as among men with high MAO activity. In this study, only the lowest and highest 10 percent of the students were compared. In both studies, a sex difference was evident. Men tended to have lower levels than did the women, and the extremely low values were found mainly in men (Buchsbaum, Coursey, and Murphy 1976).

The scientists do not yet know the basis for this sex difference. The same group, however, has conducted other studies of platelet MAO which suggest sex steroid effects on the enzyme activity. Findings such as the small but significant changes in platelet MAO during the human menstrual cycle and the higher activity in male animals during the nonmating season as compared with the mating season both have important implications.

To back up the human studies, NIMH-supported research on monkeys is furnishing preliminary evidence which links MAO activity to behavioral traits. High MAO activity correlates quite strongly with a measure called "time spent alone" and less strongly but still significantly with inactivity and passivity. Low MAO, by contrast, is associated with physical movement, social contact, and, in males, with playing.

All in all, Dr. Murphy maintains that MAO activity does indeed reflect not only a vulnerability to psychopathology but some aspects of normal behavior in humans and lower primates as well.

MAO activity level is genetically influenced. Twin studies have demonstrated a heritable component in the functioning of this particular enzyme. It would be expected anyway, since most enzymes that function in degradation pathways are under considerable genetic control. But if heredity affects the process, so does the environment. Virtually all the chemical substances known to be involved in human behavior are influenced by both genetic background and life experiences.

Transmitter levels, for instance, may be highly responsive to ongoing life experiences, such as stress. In one important test with mice, Axelrod demonstrated that living stresses could raise three- or four-fold the levels of enzymes regulating transmitter synthesis. The increases were maintained for a substantial period of time and were brought about by frequent dominance-subordinance conflicts and heavy social stimulation (Axelrod et al. 1970). One can imagine the human brain responding similarly, pouring out transmitter chemicals to contend with the increased stimulation and conflict of modern life.

Scientists all over the world are looking at the transmitters and their roles alone, together, and in different parts of the brain and body. Researchers from 29 countries convened in May 1973 at the Third International Catecholamine Symposium, held at the University of Strasbourg, France, where they presented papers on a wide range of studies, both human and animal. The Proceedings reported findings that are full of promise (Usdin and Snyder 1973).

A FEW PATHWAYS ILLUMINED

The biochemical pathways to human behavior have been only dimly seen, but they are being revealed more and more clearly. If the present course of investigation fulfills its promise, scientists should begin to tie in various mental and emotional disturbances with chemical imbalances in specific regions of the brain. They are beginning to define several types of depression and schizophrenia, separating out chemical subgroups that are biologically distinct from each other. They may even be capable of tracing familial patterns in the central nervous system that make one family susceptible to neuroticism, for instance, while another is vulnerable to schizophrenia, or paranoia, or depression. These are not predictions, but illustrations of potential, since it is impossible to say in advance what research may produce. But the potential is there, that one day knowledge of the genetic-biological foundations of behavior will allow researchers to design far more specific and effective therapy for emotional distress than has ever before been available.

REFERENCES

Axelrod, J.; Mueller, R.; Henry, J.P.; and Stephens, P.M. Biosynthesis and metabolism of noradrenaline and adrenalin after social stimulation. *Nature,* 225:1059–1060, 1970.

Baldessarini, R.J. The basis for the amine hypothesis in affective disorders. *Archives of General Psychiatry,* 32:1087–1092, 1975.

Barchas, J.D.; Ciaranello, R.D.; Dominic, J.A.; Deguchi, T.; Orenberg, E.K.; Renson, J.; and Kessler, S. Genetic aspects of monoamine mechanisms. In: Usdin, E., ed. *Neuropsychopharmacology of Monoamines and Their Regulatory Enzymes.* New York: Raven Press, 1974a. pp. 195–204.

Barchas, J.D.; Ciaranello, R.D.; Dominic, J.A.; Deguchi, T.; Orenberg, E.K.; Renson, J.; and Kessler, S. Genetic differences in mechanisms involving neuroregulators. *Journal of Psychiatric Research,* 11:360–374, 1974b.

Barchas, J.D.; Ciaranello, R.D.; Kessler, S.; and Hamburg, D.A. Genetic aspects of catecholamine synthesis. In: Fieve, R.R.; Rosenthal, D.; and Brill, H., eds. *Genetic Research in Psychiatry.* Baltimore: Johns Hopkins University Press, 1975.

Barchas, J.D.; Ciaranello, R.D.; Stolk, J.M.; Brodie, H.K.H.; and Hamburg, D.A. Biogenic amines and behavior. In: Levine, S., ed. *Hormones and Behavior.* New York: Academic Press, 1972. pp. 235–329.

Bentley, D., and Hoy, R.R. The neurobiology of cricket song. *Scientific American*, 231(2):34–52, 1974.

Benzer, S. Genetic dissection of behavior. *Scientific American*, 229:24–37, 1973.

Bodmer, W., and Cavalli-Sforza, L. *Genetics, Evolution and Man*. San Francisco: Freeman, 1975.

Breese, G.R.; Cooper, B.R.; and Hollister, A.A. Relationship of biogenic amines to behavior. In: Matthysse, S.W., and Kety, S.S., eds. *Catecholamines and Schizophrenia*. Oxford: Pergamon Press, 1975.

Broadhurst, P.L.; Fulker, D.; and Wilcock, J. Behavioral genetics. *Annual Review of Psychology*, Vol. 25. Palo Alto: Annual Reviews, 1974.

Buchsbaum, M.S.; Coursey, R.D.; and Murphy, D.L. The biochemical high-risk paradigm: Behavioral and familial correlates of low platelet monoamine oxidase activity. *Science*, 194(4262):339–341, 1976.

Buchsbaum, M.S.; Landau, S.; Murphy, D.L.; and Goodwin, F. Average evoked responses in bipolar and unipolar affective disorders: Relationship to sex, age of onset, and monamine oxidase. *Biological Psychiatry*, 7(3):199-211, 1973.

Davis, J.M. Catecholamines and psychosis. In: Friedhoff, A.J., ed. *Catecholamines and Behavior*, Vol. II, Neuropsychopharmacology. New York: Plenum Press, 1975.

Dobzhansky, T. Genetics and the diversity of behavior. *American Psychologist*, June:523–530, 1972.

Grouse, L.; Omenn, G.; and McCarthy, B. Studies by DNA-RNA hybridization of transcriptional diversity in the human brain. *Journal of Neurochemistry*, 20:1063–1073, 1973.

Hamburg, D., et al. Anger and depression in perspective of behavioral biology. In: Levi, L., ed. *Emotions: Their Parameters and Measurement*. New York: Raven Press, 1975.

Holman, B.; Elliott, G.R.; and Barchas, J.D. Neuroregulators and sleep mechanisms. *Annual Review of Medicine: Selected Topics in the Clinical Sciences*, Vol. 26. Palo Alto: Annual Reviews, 1975. pp. 499–520.

Kessler, S.; Ciaranello, R.; Shire, J.; and Barchas, J.D. Genetic variation in activity of enzymes involved in synthesis of catecholamines. *Proceedings, National Academy of Science*, 69(9):2448–2450, 1972.

Kessler, S.; Harmatz, P.; and Gerling, S.A. The genetics of pheromonally mediated aggression in mice. *Behavior Genetics*, 5(3):233–238, 1975.

Lindzey, G.; Loehlin, J.; Manosevitz, M.; and Thiessen, D. Behavioral genetics. *Annual Review of Psychology*, Vol. 22. Palo Alto: Annual Reviews, 1971.

McClearn, G., and Meredith, W. Behavioral genetics. *Annual Review of Psychology*, Vol. 17. Palo Alto: Annual Reviews, 1966.

Murphy, D.L. The behavioral toxicity of monoamine oxidase-inhibiting antidepressants. *Advances in Pharmacology and Chemotherapy*, 14:71-105, 1977.

Murphy, D.L.; Belmaker, R.H.; Buchsbaum, M.; Martin, N.F.; Ciaranello, R.; and Wyatt, R.J. Biogenic amine-related enzymes and personality variations in normals. *Psychological Medicine*, 7:149-157, 1977.

Murphy, D.L.; Belmaker, R.H.; and Wyatt, R.J. Monoamine oxidase in schizophrenia and other behavioral disorders. *Journal of Psychiatric Research*, 11:221-247, 1974.

Murphy, D.L.; Goodwin, F.; Brodie, K.; and Bunney, W. L-dopa, dopamine, and hypomania. *American Journal of Psychiatry,* 130(1):79-82, 1973.

Omenn, G. Neurochemistry and behavior in man. *Western Journal of Medicine,* 125(6):434-451, 1976.

Orten, J.M., and Neuhaus, O.W. *Human Biochemistry.* St. Louis: C.V. Mosby, 1975. 9th Ed.

Post, R.; Kotin, J.; Goodwin, F.; and Gordon, E. Psychomotor activity and cerebrospinal fluid amine metabolites in affective illness. *American Journal of Psychiatry,* 131(1):67-78, 1973.

Potkin, S.G.; Cannon, H.E.; Murphy, D.L; and Wyatt, R.J. Are paranoid schizophrenics biologically different from other schizophrenics? *The New England Journal of Medicine,* 298(2):61-66, 1978.

Reis, D.J. Consideration of some problems encountered in relating specific neurotransmitters to specific behaviors or disease. In: Matthysse, S.W., and Kety, S.S. eds. *Catecholamines and Schizophrenia.* Oxford: Pergamon Press, 1975.

Schildkraut, J. Reduced platelet monoamine activity in a subgroup of schizophrenic patients. *American Journal of Psychiatry,* 113(4):437-439, 1976.

Schlesinger, K. Genetics of audiogenic seizures, III. *Life Sciences,* 9(1): 721-729, 1970.

Skeen, J., and Thiessen, D. Paper presented at Sixth Annual Meeting, Behavior Genetics Association, Boulder, Colorado, 1976.

Slater, E., and Cowie, V. *The Genetics of Mental Disorders.* London: Oxford University Press, 1970.

Snyder, S.A. Amphetamine psychosis: A model schizophrenia mediated by catecholamines. *American Journal of Psychiatry,* 130(1):61-66, 1973.

Stein, L., and Wise, D.D. Amphetamine and noradrenergic reward pathways. In: Usdin, E., and Snyder, S.H., eds. *Frontiers in Catecholamine Research.* Oxford: Pergamon Press, 1973.

Thiessen, D. Footholds for survival. *The American Scientist,* 61(3):346-351, 1973.

Thiessen, D. A move toward species specific analysis in behavior genetics. *Behavior Genetics,* 2(2/3):115-125, 1972.

Thiessen, D. Philosophy and method in behavior genetics: Its relation to biology. In: Gilgren, A., ed. *Contemporary Scientific Psychology.* New York: Academic Press, 1970.

Usdin, E., and Snyder, S.H. eds. *Frontiers in Catecholamine Research.* Oxford: Pergamon Press, 1973.

Weiss, J.M.; McEwen, B.S.; Silva, M.T.A.; and Kalkert, M.F. Pituitary adrenal influences on fear responding. *Science,* 163(3863):197-199, 1969.

Wyatt, R.J.; Murphy, D.L.; Belmaker, R.; Cohen, S.; Donnelly, C.H.; and Pollin, W. Reduced monoamine oxidase activity in platelets: A possible genetic marker for vulnerability to schizophrenia. *Science,* 173:916-918, 1973.

Section III.

MENTAL DISORDERS

Chapter V. Schizophrenia

Iris was slow and not very competent. People in her office called her a "jerk" and she avoided them. She also avoided her father, a mean, unpredictable, critical man who drank to excess and possessed not a whit of compassion. At the age of 22, Iris broke down and the critical world began to speak through her own mind. This time when the voices of her office workers nagged and criticized, she couldn't turn them off because they were products of her own imagination. Iris had become schizophrenic.

Nora followed her sister Iris into the hospital, where hallucinations began after an unpleasant involvement with a man. Nora felt unclean—her father used to fondle her breasts—and when she first hallucinated she heard God's voice and saw demons surrounded by fire.

Hester, the third sister, broke down 2 years later at the age of 24, screaming that a wall of fire was going to consume them all.

Myra, the fourth sister, never did lose control, although periodically she goes through some very shaky periods. She is married, has two sons and keeps her distance from the family.

These are the Genain quadruplets, unique because of the rarity of monozygotic quads, and unique in psychiatric history for the appearance of schizophrenia in all four. Such an event occurs with extreme rarity, perhaps once in one to two billion births, according to Dr. David Rosenthal, at the NIMH Laboratory of Psychology, who has followed the sisters with treatments and evaluation for some 20 years.

The quads have the same genes. Thus, these women all share whatever genetic factor, or factors, may predispose humans to schizophrenia. They also share a common family environment, and there seems little doubt that this family environment, controlled by a suspicious and cruel man, was not a healthy one.

The quality of their family life is suggested by the fact that the father would sit in the house and drink behind drawn blinds, refusing to let the girls play with other children. He also patrolled the house with a gun. Sexual repression and confusion were intense. Two of the girls, caught masturbating, were clitorectomized and their hands were tied to the beds for 30 nights.

Both heredity and environment were clearly working against the Genain women. What role the genes alone would have played in causing their mental illness is impossible to say. Perhaps, if they had grown up in a better environment, not all of the women would have developed symptoms of schizophrenia. That they did suggests a much higher vulnerability than is usually evident in studies of identical twins.

The fact that one escaped serious handicap, however, demonstrates that, even under the worst circumstances, healthy forces can override the predisposing effects of genes and family. Why Myra escaped is difficult to know. Dr. Rosenthal believes that a crucial factor was her ability to maintain more distance from the family group, especially her father, and to make and keep friends on the outside. She was not happy in the family and not as caught up in it as were her three sisters. She made several attempts to break away as a teenager, and later, when it appeared that she was having a pyschotic break, she talked her family out of committing her to a hospital.

There was one other crucial factor operating for Myra, a factor that has turned up consistently in studies of identical twins at NIMH: differential treatment by the parents who tend to view one child as weaker than the other, both physically and pyschologically (Stabenau 1973). The "stronger" twin is thus imbued in the parents' eyes with more positive values and represents the parents' "good" side, while the smaller, weaker twin represents the "bad" or helpless side.

In the Genain family, Myra and Nora were considered the stronger two, and in fact, Nora was the strongest and the brightest of the four as a child. But her leadership status was apparently her downfall, because she became enmeshed too deeply in family dynamics and never broke out. Only Myra escaped. She alone is working at a job in their fourth decade of life.

Searches for a Definition

The Genain story is one of the most dramatic in the long search for the causes of pychopathology and its incidence among members of the same family. There have been "pedigree studies," largely to learn whether the schizophrenic trait is distributed along Mendelian lines, and whether the gene is dominant, recessive, or sex-linked—a type of search somewhat reminiscent of the ancient ascription to heredity "unto the third and fourth generations." Successive research studies—twin, family, adoption, biochemical—have continued the search, one of continuing challenge, to establish a definition for the illness, or illnesses, and to learn the extent of risk to family members.

A risk for what? How is concordance for schizophrenia demonstrated? Geneticists, therapists, diagnosticians agree in their disagreement about what schizophrenia really is. Finding the genetic pathway to the cricket's song was tracking to a finite end, a known behavioral goal. Finding the genetic pathway, or possibly pathways, for disordered behaviors of almost infinite variety may be rendered much more possible when those behaviors are more codified.

The diagnosis of schizophrenia is an art. Some pyschiatrists can detect the condition but they cannot validate their diagnosis, and there is not a consistent set of symptoms that occur in all schizophrenics and only in schizophrenics. Labeling these many possible symptoms is sure to engender professional disagreement. Indeed, validating criteria at present depend on genetic studies, distinguishing pharmacotherapy, and some attributes of the course of the illness.

The popular understanding of schizophrenia as a single disease is misguided. It is better understood as many symptoms in search of several diseases. "Core" schizophrenia, alone, disregarding "spectrum" and "schizoid" diagnoses, covers a broad array of mental and emotional responses which may include hallucinations and delusions, but more often does not. These behaviors, moreover, are not diseases in themselves, but manifestations of some underlying processes that assume a bewildering variety of forms.

What is schizophrenia? The question has multiple answers, depending on the needs and backgrounds of the people using the concept. Nevertheless, the American members of an international pilot study of more than 1,000 patients in nine countries did find some common ground in the kinds of behaviors that are diagnosed as schizophrenia. According to their report (Strauss, Carpenter, and Bartko 1974, pp. 62-69), the various schools of psychiatry have accepted the following major types of behavior as indicative of schizophrenia, although evidencing every symptom is not necessary for the diagnosis.

First and most widely accepted are disorders of thought processes, a loss of connectedness between thoughts, and, in addition, delusions and auditory hallucinations. Visual hallucinations are not as common in schizophrenia as is hearing voices. Delusions—firmly held false beliefs—are more difficult to diagnose because the credibility of any story depends on the social and cultural setting of the people involved. When most people believed in witches, for instance, a person could not be viewed as schizophrenic for believing his neighbor to be a witch. And in the age of telephone taps, those who think they are under surveillance may not be deluded. But the bizarre tales of the schizophrenic thought process have elements that seem immediately suspect and contrary to all evidence.

Disorders of emotional expression rank with thought disorders and delusions as critical signs. Schizophrenics may laugh when others would cry or cry when others would laugh. But particularly important is the absence of nearly all emotional expression, called "blunted affect." Lack of emotion is especially symptomatic of schizophrenia when the patient recounts in a calm, matter-of-fact way an incredible story which, if true, would cause anyone else great anger, pain, or anxiety.

The third category, disorders in personal relationships and curtailment of interpersonal relatedness, is not difficult to observe in the chronic schizophrenic, in therapy, or in other settings.

Disorders in speech, although not common, are more specific signs of schizophrenia. Among the linguistic symptoms are neologisms—new words with meaning only to the patient—and, more commonly, a flow of disordered, or incoherent words, which also show thought disorder.

Motor disorders include stereotyped, repetitive motions and a catatonic behavior called "waxy flexibility," in which the patient fails to move his arm or leg from a position someone else has placed it in.

For many years, it was apparent that schizophrenia was diagnosed twice as often in New York as in London. Why? What forces might produce twice as many schizophrenics on one side of the Atlantic as on the other? If the cultures were more dissimilar the differences could conceivably be based on a real disparity in behavior. But a 10-year-old cross-national study funded by NIMH found no real differences between several hundred New York and London patients. The behavior was the same; the psychiatric diagnosis was not, for several reasons.

There were biases that influenced diagnosis quite apart from the symptoms of the patients. In London, psychiatrists were more likely to see men than women as schizophrenic, although both sexes showed the same psychopathology. In New York psychiatrists were more likely to diagnose schizophrenia in black patients, although the symptoms and incidence were no different in black than in white patients. In both cities, younger patients were more likely to be deemed schizophrenic than were older persons with similar symptoms. Social class seemed to have little bearing—probably an insignificant result, since the patient population in the study was not drawn from a wide social range.

In both countries, patients with such obvious symptoms as delusions of control, blunting of affect, and speech disorganizations were thought to be schizophrenic. This indicates that there is a core of behaviors recognized as schizophrenic in both countries, although the actual diagnosis draws in a much wider circle of patients, many of whom may not show any of the core behaviors.

Problems with diagnosis are critical to genetic studies of schizophrenia because inconsistencies and variability have scarred the basic data on which genetic estimates are made. The question of who is schizophrenic and who is not has made scientific analysis all but impossible, unless the patients are very carefully picked. Even then they may not constitute a single homogeneous group.

Each human brain has a characteristic way of becoming psychotic. While loss of contact with reality may take the same form throughout the world, that event is not in itself a disease. It is a symptom which is frequently temporary. What actually is going on in the brain or which genes lead some people into more permanent incapacity remains unknown.

Twin, Family, and Adoption Studies

Twin studies have long been considered a prime method of confirming the input of genes to the condition known as schizophrenia—or, for that matter, to intelligence, personality, or the affective disorders. This is not because twins are more likely to be schizophrenic—nor more intelligent, more personable, or more depressed than single-birth members of the population. But twin studies help us to isolate the genetic contribution to the disease. As the co-investigators of long-term twin studies conducted at London's Maudsley Hospital put it, reviewing the behavioral genetic perspective from various recent studies of schizophrenia: "The classical comparison of MZ and DZ co-twin concordance rates assumes that within-family environmental factors are controlled and, by permitting gene dosage to vary, permits the confirmation or refutation of certain genetic hypotheses" (Gottesman and Shields 1976, p. 371).

"Concordance," or agreement, as used here, is the measure of how often *both* schizophrenic twins are diagnosed similarly. The genetic hypothesis virtually predicts concordance for monozygotic (MZ) twins since they are genetically identical; concordance in dizygotic (DZ) twins need not be identical, since they share only half their genes, on the average, just like any other brothers and sisters. On the whole, identical twins are usually considered three to five times more likely than fraternal twins to be similar in mental illness, the difference presumably revealing the impact of similar genes.

Concordance figures on schizophrenia represent a rough average of many studies over the years, and they hide an enormous variability in individual research reports. Thirty years ago, for instance, scientists were reporting concordance rates of almost 80 percent among identical twins, but with improved diagnosis and better methods of study, the rate of concordance declined. It seems that scientists saw much more schizophrenia in MZ twins in the early years, perhaps because they tended to find their subjects through mental hospitals. In addition, the question of mono- or dizygosity was not always determined with accuracy and the period of risk—and therefore of age correction in making tabulations—was not always considered. In addition, often the co-twin's diagnosis is undetermined. All of this may well have made for inflated concordance rates.

Indeed, interpretation of "concordance" has given researchers a great deal of trouble. Does it mean occurrence of the same condition, to greater or lesser extent, i.e., schizophrenia in one twin and doubtful or "borderline" schizophrenia or a schizoid condition in the other? Can there be concordance in some symptoms and discordance in others? May one twin be schizophrenic and the co-twin show some other psychopathology such as neurosis or mental retardation? Does the criterion of concordance extend only to hospital admission? Another question is whether it refers to the proportion of cases with an affected partner or the proportion of pairs in which both twins are affected.

In the sixties, new methods were applied to the problem. MZ twins were examined for concordance in various aspects of the illness and for discordance in others, the former because these were felt to represent hereditary influences and the latter probable environmental bases.

Despite these *caveats* about some of the earlier studies, they are a rich resource and an impressive legacy to the long history of twin research. Concurrently, in Europe and the United States, the studies and comparisons of data went on. The first major study, reported in 1928 by Hans Luxenberger of Bavaria, and significant because it was composed of a more systematic twin sample than most previous studies had been, showed a concordance of 59 percent in 17 pairs of MZ twins. Interestingly, 6 years later, this same researcher reported the results of another twin study: 33 percent concordance in 27 pairs of MZs. Other landmark studies previously mentioned are the Franz Kallman study of 1946 and Eliot Slater's study in Great Britain in 1953, with its reported concordance of 65 percent in 37 MZ pairs and 11 percent in 112 DZ pairs.

Three modern Scandinavian studies of Einar Kringlen have aroused a great deal of interest. The account of his research experiences is fascinating in itself; the methods attracted scholarly interest because of the attention which was given to the "pairwise" concordance method, derived from the percentage of concordant pairs in a twin population. Concordance was established not only for schizophrenia but for all types of functional psychosis and for schizophrenia and schizophreniform psychosis. Kringlen found the concordance figures for schizophrenia to be 25 to 38 percent in monozygotics and 4 to 10 percent in dizygotics, depending on whether the rates were based on registered hospital cases or on personal investigations, and on whether a wide or strict concept of the disorder was used.

Highly significant was Kringlen's classification of the co-twins of MZ index cases with typical schizophrenia. Thirty-one percent showed the same condition, typical schizophrenia, and 31 percent were classified as normal. The remaining 38 percent, for a total of 45 co-twins, showed 2 percent, reactive psychosis; 7 percent, borderline states; and 29 percent, neurosis. Another aspect of this study which presaged future research was his consideration of the premorbid personalities of future schizophrenics.

Dr. Kringlen came to the conclusion that twin investigations could not reveal anything about the mode of inheritance but, if combined with family studies, they could yield information which, to him, suggested that genetic predisposition to schizophrenia was of a polygenic nature. He felt that this best explained the existence of the full spectrum from normality to severe mental illness found in twin pairs, their siblings, and their parents. If this view of polygenic origin and of the importance of environmental factors in the etiology of schizophrenia was correct, he felt that solving the enigma of schizophrenia was unlikely to come from "any simple biochemical breakthrough," but that further meticulous research in the social sciences would provide more of the answer (1967).

In London, at about the same time, the Maudsley Hospital study proceeded from a large and valid base, a register of all twins drawn from about 45,000 consecutive admissions to both inpatient and outpatient sources, largely from the latter. The register was begun in 1948 at the instigation of Dr. Eliot Slater by asking, as part of the admissions procedure, whether the patient had a twin. In 1972, Dr. Irving Gottesman, Professor of Psychology at the University of Minnesota, and the late Dr. James Shields, from the Institute of Psychiatry (Genetics) in London, published the results of their study, with case histories of 114 twins, both MZ and DZ pairs. Their work added several new dimensions to twin study, partially because of the extensive psychometric assessments drawn from a number of psychopathic rating and personality scales and the use of blindfolded diagnoses by a six-judge panel. The careful case studies provide a rich resource into the minds and habits of schizophrenics of many degrees of illness (Gottesman and Shields 1972).

Gottesman's and Shields' computations of concordance were presented on two levels: The pairwise range of 40–50 percent for monozygotics and 9 to 10 percent for dizygotics represents the investigators' reports; the probandwise concordance, using the criterion established by their panel of judges, of schizophrenia, probable schizophrenia, and functional psychosis with schizophrenic-like features, was 58 percent for monozygotic twins and 9 to 10 percent for the dizygytics. Put another way, "Briefly, the pairwise rate expresses the degree of concordance as the percentage of all pairs in which both twins are schizophrenic, given a specified sample of twin pairs with at least one twin schizophrenic. The probandwise rate is the percentage of independently ascertained schizophrenic twins (the probands) who have a schizophrenic co-twin" (1976, p. 372).

All of this presents even finer grist for the twin-study mill. Gottesman and Shields continued using the expected and observed rates of various conditions in co-twins as a basis for studying other manifestations such as schizoidia and the schizophrenic spectrum. They see discordant MZ twins as very useful for learning about specific stressors which may precipitate schizophrenia and as an indication of the biochemical and physiological endophenotypes that are necessary for the disease or contribute to its etiology, although they are not the result of the disease (1976).

Family studies offer a different perspective on schizophrenia. In the main, they provide risk figures for mental illness by degree of relationship to the schizophrenic individual. As with the concordance rates on identical twins, the risk figures vary widely with different standards of diagnosis. Quite high risks, for instance, are given by Fuller and Thompson (1960), who estimate the risk to children of one schizophrenic parent at 16 percent. With two schizophrenic parents, they estimate that 35 to 68 percent of the children will also become ill. The risk for siblings is 11 to 14 percent and for grandchildren, 4 percent.

Most contemporary estimates, however, indicate a lower risk for the family members of affected individuals. Twenty-five studies summarized by Zerbin-Rudin, as represented by Slater and Cowie (1971), put the risk to children at 12 percent, and the risk to siblings at 8 to 14 percent, depending on whether the parents show schizophrenia. Grandchildren have a risk somewhat less than 3 percent.

While the International Pilot Study searched for ways to define schizophrenia, teams in the United States searched for and found a genetic component for the condition. The questions raised by these simultaneous searches ache for answers. Will the two levels meet and where? When and if the genetically influenced behavior is ever defined, will it be anything now known as schizophrenia? It is obvious that the genetic contribution cannot be deduced solely from family studies, nor from twin studies alone. The skein of each individual life is entangled with genetic and socioenvironmental factors, biochemical actions and reactions, and family interactions all wound together. To separate the effects of rearing and environment from the role of genes, it was decided to study adopted children, comparing illness rates in the biological as compared to the adopted families.

The first study, reported by Leonard Heston in 1966, traced the fates of 47 infants born to schizophrenic mothers and placed in foundling homes or foster families. At the time, the prevailing belief in psychiatry was that mothers caused the sickness in their children through abnormal rearing of some kind. So it came as a shock to discover in 1966 that even when the children were taken from their mothers in the first few days of life, they still became schizophrenic at the expected rate for offspring of such parents. By comparison, a control group of adoptees, who grew up in the same type of foster care institutions and foster homes, but who were born of nonschizophrenic mothers, had no schizophrenia. Where was the effect of rearing, if the offspring of schizophrenics became sick anyway?

But that was not the only surprise to come from this richly productive study. Heston also found that half of the experimental group (26 of 47) were in real psychosocial trouble, although only five could be called schizophrenic. Of 30 men in the experimental group, 7 were felons, 8 had been discharged from the armed services for behavioral or psychiatric reasons, and another 3 were rejected by the services for the same reasons. Several had an antisocial personality of an impulsive nature, characterized by poorly planned crimes and alcohol abuse. Heston (1966, p. 824) described such individuals this way: They "tended to live alone—only one was married—in deteriorated hotels and rooming houses in large cities. . . . They worked at irregular casual jobs such as dishwasher, race-track tout, parking attendants. When interviewed they did not acknowledge or exhibit evidence of anxiety. Usually secretive about their own life and circumstances, they expressed very definite though general opinions regarding social and political ills. In spite of their suggestive life histories, no evidence of schizophrenia was elicited in interviews. No similar personalities were found among the control subjects."

A second group of the experimental subjects, mostly women, showed unstable personalities characterized by anxiety, hyperirritability, and depression. Whether this type of emotional behavior is related to schizophrenia or not is causing much current controversy. Many believe that anxiety and panic are signs of depressive disorder rather than schizophrenia, and it is possible that some of the mothers in this study would have been diagnosed as having manic-depressive psychosis rather than schizophrenia, if evaluated today.

Three behavioral traits were found almost exclusively in the experimental group—musical ability, strong religious expressions, and problem drinking. A surprising finding, in many ways, was the behavior of this other half of the experimental group—the people who were not psychologically disabled. They were, by contrast, rather successful adults, above normal when compared to the control group.

"They were not only successful adults but in comparison to the control group were more spontaneous when interviewed and had more colorful life histories. They held more creative jobs: musician, teacher, home designer; and followed the more imaginative hobbies: oil painting, music, antique aircraft. Within the experimental group there was much more variability of personality and behavior in all social dimensions.

"That the genetic endowment in families with schizophrenia may contain the potential for supernormal development is firmly believed by some investigators and disbelieved by others. The issue is complex and there is little evidence with which to resolve it."

Heston's study traced schizophrenia from mother to offspring. What would happen if the subject were turned around, starting with a group of adopted schizophrenic individuals and tracking down their parents?

The first reports on the second of these studies were made at a conference at Dorado Beach, Puerto Rico, in 1967 and left no doubt that schizophrenia is inherited to an important degree. This elegantly constructed study design of schizophrenic adults in Denmark who had been adopted as infants remains a model in the field of behavior genetics. The study was designed to test a genetic component in chronic schizophrenia and, later, schizophrenia spectrum disorders, the hypothesis being that, if schizophrenia runs in families because of genes, the illness should be concentrated among biological relatives.

Drs. Seymour Kety, David Rosenthal, and Paul Wender wanted to do their adoption studies in the United States, but it soon became clear that would be impossible. The plethora of different adoption agencies, different policies, different records would defeat any scientific effort, not to mention the monumental problem of trying to trace biological relatives in a country where people migrate widely. This was in 1961.

There was one country, Denmark, where every adopted child in the nation was entered into one register along with the names of the adoptive parents, the biological mother and the putative father. The adoptee tended to be illegitimate and, in a genetic study, proof of paternity could be a serious problem. But in Denmark, fathers had to show up at the adoption agency, claim paternity and contribute to adoptive costs. So it was assumed by the NIMH scientists that those men who did show up at least thought they were the fathers.

In addition to these records, Denmark also keeps a national pyschiatric register with the names and diagnoses of 95 percent of all Danes who have ever seen a psychiatrist or been admitted to a psychiatric hospital or facility. These people are not all mentally ill. The records simply reflect a history of having sought mental help for whatever reason, from schizophrenia to problems in living.

Finally, Denmark maintains a population register called the Folkeregister which contains the name, birth date, and address of every individual who has ever lived in Denmark for more than a few weeks. Records of addresses are kept from birth and contain the names of other members in the household as well.

Thus, from records alone, the scientists could trace the adopted children and psychiatric histories. They could also trace both lines of relatives—one on the biological side, one on the adopted side—and their psychiatric histories. Everything was present for finding out which side produced most of the schizophrenia—for disentangling the variables of environment and genes.

From a total of 5,500 children adopted in Copenhagen between the years 1927 and 1948, the team, which also included a Danish psychiatrist, Dr. Fini Schulsinger, found 33 who had become schizophrenic in adulthood. These became the index cases, the point from which the genetic search would begin. Next the team selected 33 mentally normal adoptees, who were like the schizophrenics in age, sex, social class, and other features, except that they had no history of any psychiatric treatment. These would form a normal baseline for the incidence of schizophrenia in the general population, against which the families of schizophrenics could be compared.

The team now had four sets of relatives to trace: a biological and adopted set for the 33 schizophrenics and a biological and adopted set for the 33 controls. If all or most of the schizophrenic behavior was concentrated in only one line—the biological relatives of the index cases—then the genetic hypothesis would be confirmed, or at least strongly supported.

Altogether there were 512 relatives—parents, siblings, and children—some related by genes, others by environment. Each relative was evaluated on the basis of psychiatric, legal, military and State welfare records. Each was placed in a category: normal . . . chronic schizophrenic . . . acute schizophrenic . . . latent schizophrenic . . . uncertain schizophrenic . . . schizoid personality . . . inadequate personality . . . manic-depressive, and so on.

The psychiatrists worked completely blind. That is, all clues as to whether the relative was biological or adopted had been removed from the records. All identifications were by code to make it as difficult as possible for hope and bias on the part of the scientists to creep into their judgment.

One problem became apparent immediately. There were not enough hospitalized schizophrenics among the relatives to make any judgments at all. Schizophrenic-like behavior appeared in the records, but without a formal diagnosis. Since Danes do not accept the latent or "borderline" schizophrenic category currently in use in the United States, a narrower group gets defined as schizophrenic in Denmark. The team made a decision, a controversial one, to expand the criteria into a new category called "schizophrenic spectrum" disorders, which could apply to people with doubtful behavior but no diagnosed manifest disease.

By now, the team members (Kety, Rosenthal, Wender, Schulsinger) were really following their noses. Precisely what behaviors should be called "schizophrenic spectrum" disorders? What qualities of mind and emotion suggest this *potential* for serious mental disorders?

Using their own long years of experience with schizophrenia as a guide, the team identified schizophrenic-like behaviors in people who had never been identified before and who would probably not be diagnosed under any other system. Time and again, the four thrashed out their disagreements over individuals and diagnoses, each time reaching a consensus: schizophrenic spectrum: yes or no.

Then they broke the code, to see whether the relative was biologically or adoptively related to the diagnosed schizophrenic. Whatever it was, or is, schizophrenic spectrum disorders were concentrated in the *biological* relatives of the index cases from whom they had been separated 25 to 50 years before.

But the team was not satisfied with this first genetic evidence. It was important to find definite schizophrenia among the relatives to be sure of the genetic link. For that, the records were no longer enough and the decision was made to interview every one of the 512 relatives who were still living and would agree to an interview. Ninety percent agreed. (Of those relatives who had died, most were also concentrated in the biological line of index cases, and the deaths frequently proved to be suicides—often a sign of mental disturbance.)

A Danish psychiatrist, Dr. Bjorn Jacobsen, undertook the interview, devoting 2 years to the job and an average of 35 pages of transcript and evaluation per individual. Again each relative was categorized by the team, which this time based their judgments on interviews and records together. Now definite schizophrenia appeared in people who had never come to the attention of the mental health establishment but who showed clear symptoms, even by the more rigorous Danish standards.

When the code was broken this time, both spectrum disorder and definite schizophrenia were concentrated in the one biological group of index cases to a remarkable degree.

Of 173 biological relatives of schizophrenic adoptees, 14 individuals, 8 percent, were schizophrenic. In the other three lines, the incidence of schizophrenia was 1 to 3 percent, normal for any population. For the broader "spectrum" category, which included definite schizophrenia, the figures were roughly doubled. Sixteen percent of the biological relatives of index cases showed traits of mind and emotion that these scientists believe are related to schizophrenia. By contrast, among the adopted relatives and relatives of control cases, the incidence of such traits reached only 4 to 6 percent.

This was a very significant difference, compelling evidence for a genetic factor in schizophrenia. Here were adopted children who became mentally ill as adults, yet the people with whom they grew up were not sick. Their genetically related relatives, with whom they had never lived, showed four to five times the usual incidence of schizophrenia.

Dr. Wender estimates that spectrum disorders are five to ten times as frequent in the normal population as schizophrenia itself. He believes these mild disorders form a common genetic component in a wide range of ordinary psychological problems. He stated that 8 to 16 million Americans (4 to 8 percent of the population) might be affected by spectrum behaviors, an enormous, biologically based mental health problem, if confirmed. These affected individuals would probably not be incapacitated, but they might be plagued with distressing problems. Many of the Danish people with mild disorders described miserable, although functional, lives. Such a person, for instance, might suffer from inordinate suspiciousness, vague or tangential thinking, eccentric behavior or social isolation—some of the behavioral cues the research team was responding to.

One of the most common traits in the spectrum was the tendency to take casual incidents in a personal way. Called "ideas of reference," this includes paranoid thinking as well as the tendency to see personal meaning in inconsequential events (as in: "The pencil is moved; my secretary is trying to signal me.") Also common were eccentric ideas—which don't make sense to the individual's peers.

In illustration, Dr. Kety described this hypothetical individual: "He is withdrawn, he has few friends. He is very strongly religious and he reads a great deal of philosophy, which is something schizophrenics often do. He has an inappropriate grin when he talks. He flunked out of school although he is obviously intelligent. He never married. He doesn't go out often and his room is very messy. He looks unkempt. He has a lower level job than his social class or intelligence would warrant."

In another instance, more rare, the individual might seem perfectly normal, but suddenly come up with a delusion. One such woman, a nursing supervisor, was convinced she was being devoured by cancer, despite all medical evidence to the contrary.

"She said in a very matter-of-fact way that this had been going on for years, and it was quite clear that she didn't have cancer. What do you do with someone like that? It was the only major sign of mental illness. Maybe there were a few other funny things," said Dr. Kety.

Except in the case of the full-blown delusion, the team rarely diagnosed people with only one symptom. Typically they responded to a constellation of traits, among which some prominent signs were fuzzy, illogical talking which seemed never to come to the point; odd ideas and lack of emotional rapport, as well as suspiciousness, social isolation, and frequent episodes of disassociation (feelings that one is playing a role) which were not linked to anxiety attacks. Anxiety was not one of the spectrum behaviors.

None of these behaviors was limited to a schizophrenic's family. They also appeared among 4 to 6 percent of the relatives of psychiatrically normal people, which suggests that in the general population as a whole, the chances are that most people showing such behaviors will not have schizophrenia in their background.

Just what this mild "schizophrenia spectrum" includes remains something of a mystery. The scientists were using their clinical judgment, not questionnaires, so the process of defining their behavioral genetic track of this spectrum is difficult. It involves codifying the behaviors, confirming the genetic results in new groups of people and finally boiling down the range of traits to some personality sets that seem to hang together and have a proven association with schizophrenia.

There is no longer any doubt that genes are clearly involved in some, possibly all forms of schizophrenia—chronic, certainly, spectrum, and acute. But how does one identify those genes, specify the enzymes, and many other factors that place people at risk, and perhaps circumvent the process?

The "diathesis-stressor framework," proposed by Irving Gottesman and James Shields, that schizophrenia is a genetic factor interacting with nonspecific, perhaps universal, environmental factors, is provocative for the twin-adoption-environment approach. Ten years ago, Dr. David Rosenthal (1968, p. 417) wrote of the two influences: "I think that Dr. Erlenmeyer-Kinling came closest to formulating the issue in a way that is compatible with current knowledge and conducive to developing productive research in the future: She says: 'The question to be asked is not: What are the relative contributions of heredity and environment? but rather What kinds of environmental input trigger manifestations of the disorder in genotypically vulnerable persons, and why are these important in a psychophysiological sense?' I myself would have phrased the same point in this way: How do the implicated hereditary and environmental variables interact or coact to make for various kinds of schizophrenic and nonschizophrenic outcomes? The level at which the analysis of interaction would be conducted, i.e., physiological, psychological, or psychophysiological, would be left to the individual investigators, but all such analyses could be informative."

The hope that any behavioral traits can lead the way may well be frustrated. Psychology is not an easy clue to genetic processes, and it remains totally unclear which schizophrenic behaviors are genetically influenced. To anyone involved in schizophrenia research, the links between behavior and genes may not seem very close at this time. The mind has myriad ways of responding to one stimulus, and a single genotype can be expressed in several ways at the level of personality and behavior.

Shields, Heston, and Gottesman (1975, p. 192) have pointed out that it is difficult to distinguish the affected from the unaffected members of a family with schizophrenia. Among the major reasons: (1) the expression of one gene can vary widely, and (2) several genes might be involved so that behavioral characteristics shade indistinguishably one into the other.

The widely varying expressions of one gene are nowhere more evident than in the rare disease called the Lesch-Nyhan syndrome. Caused by one defective enzyme, the syndrome leads to a severely painful behavior that includes self-mutilation, mental retardation, and neuromuscular disorders. In people suffering from this condition, the defective enzyme (hypoxanthine quanine phosphoribosyltransferase) shows an extraordinarily low rate of activity—.005 percent of normal in red blood cells. In other family members, however, the rate of activity varies up to 1 percent of normal and is associated with different consequences. Enzyme activity in the range of 0.01 percent to 0.5 percent of normal occurs in family members with various levels of retardation and neuromuscular disorders. Activity of about 1 percent of normal is associated with gout.

"The main point is a simple one. There would be no possible way on clinical grounds to group all of the clinical disorders associated with deficiencies in the activity of this one enzyme into one clinical syndrome, not even those disorders appearing in one family." (Shields, Heston, and Gottesman, p. 193)

Possibly several alleles (contrasting traits) of one gene are responsible for different enzyme levels. On the other hand, variation can also be attributed to environmental effects on gene expression and it may be influenced by the undetected impact of gene interaction. Other loci on the chromosome and their enzyme products can alter the activity of any one gene, thus causing variable levels in deficiency. Consequently, even if one gene alone were found at the base of schizophrenia—which few people believe is the case—its expression could take many different forms.

Perhaps more difficult is the attempt to make behavioral distinctions when many genes are involved. This difficulty presupposes the polygenic hypothesis of schizophrenia, which postulates that the combined action of many genes is responsible for schizophrenic behavior. A polygenic system makes discrete categories inappropriate because when many genes influence one type of behavior they produce continuous variation, ranging from no trait at all to prominent evidence of it. In this case, either/or distinctions about the presence or absence of a trait are especially misleading because the truth lies in degrees. Intelligence is thought to be a product of a polygenic system. So is height. To dichotomize the population into tall people and short people or smart people and stupid people requires that some arbitrary division be made in what is actually a continuum.

If schizophrenia is based on a polygenic system, then the relevant genes may exist in continuous variation among a large segment of the general population. Appearance of disease would then be affected by the number of relevant genes inherited, the "genetic loading" for schizophrenia. Some individuals, for instance, would carry high risks for developing severe disease, because they have all the important genes underlying this type of behavior. Others would inherit only a few genes and possibly show a few mental oddities but carry relatively low risks for developing diagnosable illness.

A recent mathematical model for such a system theorized that 1 out of 11 individuals in the population at risk for schizophrenia would carry an extremely high genetic risk for disease—99 percent. To these researchers, this indicates that those at the extreme end of the continuum with an overwhelming genetic load would have little chance of avoiding illness (Matthysse and Kidd 1976). The authors concluded their analysis by saying, however, that neither the single gene hypothesis or the polygenic system actually fits existing data on the incidence of schizophrenia. The lack of fit is not surprising in view of the impact that environment and life stresses may be expected to have on the appearance of disease.

BIOCHEMICAL AND NEUROPHYSIOLOGICAL STUDIES

A common opinion among many leading investigators in schizophrenia is that both the monogenic and polygenic hypotheses are correct. In the final accounting, it may well be found that some people with schizophrenia have a single defective enzyme, while others suffer from a more generalized biological deficit mediated through a polygenic system.

The study of psychological traits, by itself, has not yet led and may never alone lead to specific genes. Other scientists are focusing more directly on the biological level. Clinical descriptions such as the spectrum disorder thus serve as a convenient way to isolate an interesting group of people for biological study.

Such a research approach has been pursued for years with agonizingly slow results. A main problem has been the difficulty in finding a homogenous group of schizophrenics; consequently scientists have not yet found consistent biological differences between schizophrenics and other people. But even though evidence of abnormal brain activity in schizophrenics is still sparse, several promising avenues of research are open.

One of these seems to be relatively simple, that victims of schizophrenia are suffering from the action of a natural (or endogenous) hallucinogen which their own bodies have produced. Similarities between an LSD trip and the mental experiences of an acutely ill, hallucinating schizophrenic are remarkable, and there was a time when LSD was considered *the* drug analogue of schizophrenia. That period is past, but it left an awareness that some hallucinogens, especially mescaline and dimethyltryptamine (DMT) are not so different in chemical structure from substances produced naturally by the brain. All that was needed was the addition of some methyl groups to an important endogenous brain amine like tryptamine and you would have DMT, a known, powerful hallucinogen.

Several years of research revealed that there was indeed an enzyme, N-methyltransferase, in the brains and bodies of both animals and man capable of transforming tryptamine into DMT. Early in the 1970s an NIMH team finally was able to show that the activity of this enzyme in blood platelets was elevated in psychotic as compared with normal subjects, a difference that was greatly diminished by dialysis. The elevation in platelets, however, was not restricted to schizophrenics, leading the researchers to believe that this abnormality is nonspecific, since it was higher than usual in depressed psychotics as well. The authors speculate that the resulting activity may be related to the stress of psychosis in general, since there was no control for this factor. They think, too, that such a substance could be related to the finding of low platelet MAO in some depressed patients and schizophrenics (Wyatt, Saavedra, and Axelrod 1973, p. 758).

In companion research, the same team also studied identical twins concordant for schizophrenia to find a potential genetic link. If the levels were the same for the two twins, although only one was sick, it would suggest a genetic cause; if the levels varied with the illness, it would indicate that this particular enzymatic action was either a result of the disease or environmental stresses related to it. The results did not support a genetic cause; the schizophrenic twins had higher mean levels of enzyme activity than the nonschizophrenic co-twins (Wyatt, Saavedra, et al. 1973, p. 1359). This does not rule out the possibility of some genetic control over the DMT-producing machinery, but it makes that lead less promising. Whether the brains, like the bloodstreams, of psychotic subjects were producing high levels of DMT remains unknown.

If some people believe LSD was the best model for schizophrenia, others found a better analogue in amphetamine psychosis. That perception has proved more productive over the years. The action of amphetamines in the brain has provided key information as to the possible underlying biochemical causes of schizophrenia. In sufficient doses, amphetamines can produce in some individuals a psychotic episode nearly indistinguishable from a paranoid schizophrenic reaction. It apparently does this by blocking the reuptake of two important neurotransmitters—dopamine and norepinephrine—causing an increased pileup of these chemicals at their neuron synapses. Dopamine in particular seems to be responsible for the psychotic effects (Taylor and Snyder 1971; Snyder 1971, p. 19).

The amphetamine hypothesis, however, has a large deficiency. Amphetamine-induced psychosis, according to many observers, is not accompanied by a thought disorder, which is used often in defining the characteristics of schizophrenia. Dr. Richard Jed Wyatt (1978) says: "As good as the amphetamine model may be, nobody has yet proposed that amphetamine is an endogenous compound or is capable of producing schizophrenia. Phenylethylamine (PEA) is an endogenous compound structurally identical to amphetamine except for the methyl group on the side chain alpha carbon. Because of this similarity, we and others have wondered whether PEA might act like an endogenous amphetamine and perhaps be responsible for some forms of schizophrenia, in particular, paranoid schizophrenia."

Dr. Wyatt and his group offer as one hypothesis the observations that PKU children who have high concentration of PEA show some signs similar to those produced by amphetamine and that some PKU adults showed psychotic episodes and were treated as schizophrenics. This theory, presented in a speech at the Second Rochester International Conference on Schizophrenia in 1976, may be another piece to be fitted into the jigsaw puzzle to help complete the picture. Certainly, it is a piece to consider.

The amphetamine work might have gone unnoticed in the beginning were it not for the action of phenothiazines, the major tranquilizers used in the treatment of schizophrenia. Phenothiazines straighten out much of the disordered thinking in schizophrenia, but because they block dopamine they also produce an unfortunate and serious side effect—muscular tremors that resemble Parkinson's disease. Efforts to find a drug to treat psychosis without also causing symptoms of Parkinsonism have not so far been successful.

Parkinson's disease is caused by a deficiency of dopamine in critical areas of the brain. That discovery, which led to the first effective treatment of Parkinsonism with L-dopa, a precursor of dopamine, was one of the great medical research advances of the 1960s.

But as a side effect of its benefits in the muscular area, L-dopa also precipitated psychotic-like responses in some patients. As with the amphetamines, a flow of dopamine was capable of producing psychotic reactions. Some basic and very important brain mechanisms were obviously involved. Drugs that blocked dopamine action acted against psychotic thought and produced tremors. Drugs that enhanced dopamine action stopped the tremors but promoted psychotic thought. Just what was going on in the brain remained a mystery. The evidence was tantalizing, but knowledge of the brain's biochemical machinery was frustratingly weak. There was no way to demonstrate that dopamine activity was actually abnormal in the brains of schizophrenics.

Another attempt to find out just what was going on in the brain led to the discovery that cadaver brains of schizophrenics showed a deficiency (by 40 percent) of an enzyme called DBH—dopamine-beta-hydroxylase (Wise and Stein 1973). This is an enzyme that converts dopamine into norepinephrine, which is directly and intimately involved in mood and emotional behavior. If the DBH deficiency were preventing an important conversion, it could lead to a buildup of dopamine and a loss of norepinephrine. That, accordingly to some thinkers, might explain the symptoms of schizophrenia.

Efforts to confirm the DBH work have not yet been successful. An NIMH team found a slight deficiency (15 percent) in autopsy studies but think it is an artifact. No other chemical abnormalities have been found in the brains of schizophrenics, dead or alive.

A third avenue of research postulates that schizophrenia may be due to an excess, not just of dopamine, but of catecholamine neurotransmitters in general. Such an excess could conceivably predispose the individual to hyperarousal and hyperactivity, due to an overly active nervous system. According to Dr. William Pollin, formerly of NIMH, such a theory of general catecholamine excess fits well with the clinical picture of a preschizophrenic as fearful, hypersensitive, and distractable. It also fits the drug information.

Dopamine-enhancing drugs are not the only ones capable of precipitating a psychosis. Antidepressant drugs have the same potential, and it is believed that these drugs fight depression by elevating catecholamine levels, particularly norepinephrine. Actually, all drugs capable of setting off a psychotic episode—amphetamines, antidepressants, L-dopa, and steroids—seem to activate several transmitter systems at once. Dr. Pollin proposes that this activation "is an important central phenomenon in the development of a variety of psychotic states."

Reason to pursue the excess catecholamine hypothesis comes from one other bit of biochemical evidence which has caught the imagination of scientists: Studies of chronic schizophrenics revealed that MAO activity was abnormally low in blood platelets. In two-thirds of the patients, levels were as low as in people given MAO inhibitors for depression. A group of acute schizophrenics, however, were normal in this respect, indicative that acute and chronic schizophrenia are not the same illness (Carpenter, Murphy, and Wyatt 1975).

More evidence for this statement appears in recent reports based on the presence of lowered platelet monoamine oxidase (MAO) activity in chronic schizophrenics. Working from the finding of the Danish adoption studies which suggests that acute and chronic schizophrenia may be different illnesses, a group of researchers at the NIMH studied reduced platelet MAO activity as a possible genetic marker for vulnerability to schizophrenia, beginning with 13 pairs of monozygotic twins discordant for schizophrenia and 23 normal twin controls. They confirmed an earlier study which indicates that some persons showing the syndrome have lower platelet MAO activities than do most normals. They also found a high correlation for this between nine pairs of normal MZ twins. Because the twins under observation at that time were discordant for the disorder yet also showed low platelet MAO activity, the conclusion was that low platelet MAO might be a marker for *vulnerability* to schizophrenia rather than a marker for the disorder itself (Wyatt, Murphy, et al. 1973).

It seems clear from this slow and agonizing search through brain, blood, and urine samples that the answers to schizophrenia will require basic advances in the neurosciences. It seems possible that clues to vulnerability might come from research findings of another approach isolating factors of invulnerability: What protects some individuals who share the same genes as a schizophrenic from being schizophrenic themselves?

There is a long way to go, but we have learned a great deal in the past decade alone. And despite differences in disciplines, techniques, and philosophy, all researchers would undoubtedly concur in the comment (Gottesman and Shields 1976, p. 389): "Schizophrenia research ought to be avoided by those individuals who cannot readily tolerate ambiguity and uncertainty."

REFERENCES

Carpenter, W.T.; Murphy, D.L.; and Wyatt, R.J. Platelet monoamine oxidase activity in acute schizophrenia. *American Journal of Psychiatry*, 132 (4):438–441, 1975.

Carpenter, W.T.; Strauss, J.S.; and Bartko, J.J. Use of signs and symptoms for the identification of schizophrenic patients. *Schizophrenia Bulletin*, No. 11, 37–49, 1974.

Fuller, J.L., and Thompson, W.R. *Behavior Genetics*. New York: Wiley, 1960.

Garelis, E.; Gillin, J.C.; and Wyatt, R.J. Elevated blood serotonin concentrations in unmedicated chronic schizophrenic patients: A preliminary study. *American Journal of Psychiatry*. 132(2):184–186, 1975.

Gottesman, I.I. Schizophrenia and genetics: Where are we? Are you sure? In: Wynne, L.; Cromwell, R.L.; and Matthysse, S., ed. *The Nature of Schizophrenia: New Approaches to Research and Treatment*. New York: Wiley, 1978.

Gottesman, I.I., and Shields, J. A critical review of recent adoption, twin, and family studies of schizophrenia: Behavioral genetics perspectives. *Schizophrenia Bulletin*, 2(3):360–401, 1976.

Gottesman, I.I., and Shields, J. *Schizophrenia and Genetics: A Twin Study Vantage Point*. New York: Academic Press, 1972.

Gottesman, I.I., and Shields, J. Schizophrenia in twins: 16 years' consecutive admissions to a psychiatric clinic. *British Journal of Psychiatry*, 112(489):809–818, 1966.

Hawk, A.B.; Carpenter, W.T.; and Strauss, J. Diagnostic criteria and five-year outcome in schizophrenia. *Archives of General Psychiatry*, 32:343–347, 1975.

Heston, L.L. Psychiatric disorders in foster home reared children of schizophrenic mothers. *British Journal of Psychiatry*, 112(489):819–825, 1966.

Heston, L.L. The genetics of schizophrenic and schizoid disease. *Science*, 167:249–255, 1970.

Heston, L.L. Schizophrenia: Genetic factors. *Hospital Practice*, June: 43–49, 1977.

Itil, T.M.; Hsu, W.; Saletu, B.; and Mednick, S. Computer EEG and auditory evoked potential investigations in children at high risk for schizophrenia. *American Journal of Psychiatry*, 131(8):892–899, 1974.

Kety, S.S. Progress toward an understanding of the biological substrates of schizophrenia. In: Fieve, R.R.; Rosenthal, D.; and Brill, H., eds. *Genetic Research in Psychiatry.* Baltimore: The Johns Hopkins University Press, 1975.

Kety, S.S.; Rosenthal, D.; Wender, P.H.; Schulsinger, F.; and Jacobsen, B. Mental illness in the biological and adoptive families of adopted individuals who have become schizophrenic: A preliminary report based on psychiatric interviews. In: Fieve, R.R.; Rosenthal, D.; and Brill, H., eds. *Genetic Research in Psychiatry.* Baltimore: The Johns Hopkins University Press, 1975.

Kidd, K.K. On the possible magnitudes of selective forces maintaining schizophrenia in the population. In: Fieve, R.R.; Rosenthal, D.; and Brill, H., eds. *Genetic Research in Psychiatry.* Baltimore: The Johns Hopkins University Press, 1975.

Kringlen, E. Heredity and social factors in schizophrenic twins: An epidemiological clinical study. In: Romano, J., ed. *The Origins of Schizophrenia.* Amsterdam: Excerpta Medica Foundation, 1967.

Matthysse, S.W. Missing links. In: Wynne, L.; Cromwell, R.L.; and Matthysse, S., eds. *The Nature of Schizophrenia: New Approaches to Research and Treatment.* New York: Wiley, 1978.

Matthysse, S.W., and Kidd, K.K. Estimating the genetic contributions to schizophrenia. *American Journal of Psychiatry,* 133(2):185-191, 1976.

NIMH Report of the United States-United Kingdom Cross National Project. *Schizophrenia Bulletin,* No. 11:29-37, 1974.

Pollin, W. The pathogenesis of schizophrenia. *Archives of General Psychiatry,* 27:29-37, 1972.

Rosenthal, D. The concept of subschizophrenic disorders. In: Fieve, R.R.; Rosenthal, D.; and Brill, H., eds. *Genetic Research in Psychiatry.* Baltimore: The Johns Hopkins University Press, 1975.

Rosenthal, D. *The Genain Quadruplets.* New York: Basic Books, 1963.

Rosenthal, D. *Genetics of Psychopathology.* New York: McGraw-Hill, 1971.

Rosenthal, D. The heredity-environment issue in schizophrenia. In: Rosenthal, D., and Kety, S. *Transmission of Schizophrenia.* London: Pergamon Press, Ltd., 1968.

Rosenthal, D. An historical and methodological review of genetic studies of schizophrenia. In: Romano, J., ed. *The Origins of Schizophrenia.* Amsterdam: Excerpta Medica Foundation, 1967.

Sartorius, N.; Shapiro, R.; and Jablensky, A. The International Pilot Study of Schizophrenia. *Schizophrenia Bulletin,* No. 11:21-34, 1974.

Sartorius, N.; Jablensky, A.; Strömgren, E.; and Shapiro, R. Validity of diagnostic concepts across cultures: A preliminary report from the International Pilot Study of Schizophrenia. In: Wynne, L.C.; Cromwell, R.L.; and Matthysse, S., eds. *The Nature of Schizophrenia: New Approaches to Research and Treatment.* New York: Wiley, 1978.

Segal, J., ed. *Research in the Service of Mental Health: Report of the Research Task Force of the National Institute of Mental Health.* Washington, D.C.: U.S. Government Printing Office, 1975. DHEW Pub. No. (ADM) 75-236.

Shields, J.; Heston, L.L.; and Gottesman, I.I. Schizophrenia and the schizoid: The problem for genetic analysis. In: Fieve, R.R.; Rosenthal, D.; and Brill, H., eds. *Genetic Research in Psychiatry.* Baltimore: Johns Hopkins University Press, 1975.

Slater, E., and Cowie, V. *The Genetics of Mental Disorders.* London: Oxford University Press, 1971.

Snyder, S.H. Catecholamines in the brain as mediators of amphetamine psychosis. *Archives of General Psychiatry,* 27:169-179, 1972.

Stabenau, J.R. Research in genetics and psychiatry. *World Journal of Psychosynthesis,* 6(4), 1974.

Stabenau, J.R. Schizophrenia: A family's projective identification. *American Journal of Psychiatry,* 130(1):19-23, 1973.

Stevens, J.R. An anatomy of schizophrenia? *Archives of General Psychiatry,* 29:177-189, 1973.

Strauss, J.S., and Carpenter, W. Characteristic symptoms and outcome in schizophrenia. *Archives of General Psychiatry,* 30:429-434, 1974.

Strauss, J.S.; Carpenter, W.; and Bartko, J.J. Speculations on the processes that underlie schizophrenic symptoms and signs. *Schizophrenia Bulletin,* No. 11, 61-71, 1974.

Taylor, K.M., and Snyder, S.H. Differential effects of d- and l-amphetamine on behavior and on catecholamines in dopamine and norepinephrine containing neurons in rat brain. *Brain Research,* 28:295-309, 1971.

Wise, C.P., and Stein, L. Dopamine-beta-hydroxylase deficits in the brains of schizophrenic patients. *Science,* 181:344-347, 1973.

Wyatt, R.J. Is there an endogenous amphetamine? A testable hypothesis for schizophrenia. In: Wynne, L.; Cromwell, R.L.; and Matthysse, S., eds. *The Nature of Schizophrenia: New Approaches to Research and Treatment.* New York: Wiley, 1978.

Wyatt, R.J.; Murphy, D.L.; Belmaker, R.; Cohen, S.; Donnelly, C.H.; and Pollin, W. Reduced monoamine oxidase activity in platelets: A possible genetic marker for vulnerability to schizophrenia. *Science,* 173:916-918, 1973.

Wyatt, R.J.; Saavedra, J.M.; and Axelrod, J. A dimethyltryptamine-forming enzyme in human blood. *American Journal of Psychiatry,* 130(7):754-760, 1973.

Wyatt, R.J.; Saavedra, J.M.; Belmaker, R.; et al. The dimethyltryptamine-forming enzyme in blood platelets: A study in monozygotic twins discordant for schizophrenia. *American Journal of Psychiatry,* 130(12):1359-1361, 1973.

Zahn, T.P. Psychophysiological concomitants of task performance in schizophrenia. In: Kietzman, M., et al., eds. *Experimental Approaches to Psychopathology.* San Francisco: Academic Press, 1975.

Chapter VI. Depression

Depression, a disturbance of mood, with its accompanying lethargy or euphoria, is probably the most often and the most poignantly described of the mental disorders. Since the ancient Greeks, and since the Biblical Job and Saul, it has been a part of man's condition. The word escapes clear definition, perhaps because of this long history and because the word can indicate either a transitory "feeling" we all experience or a serious debilitating illness.

Clinicians are still seeking agreement as to causes: biochemical, genetic, psychological, social, or, more likely, a combination of these factors; as to epidemiology and linkages: sex, age, or family; and as to diagnosis: neurotic, unipolar, bipolar, or spectrum. They are seeking too, for the most effective therapy: antidepressants, electroshock, psychotherapy, or behavior therapy. These are only a few selections from the panoply of approaches to depression, but with new research tools, new drug treatments, and other new therapeutic agents, bits and pieces are falling into place. The nature and etiology of this condition, sometimes nominated in this country as the "mental illness of the 1970s," are becoming better understood and, therefore, more susceptible to treatment.

Like schizophrenia, depression comprises a large, amorphous group of clinical symptoms that probably represent several biological subgroups. The aim of scientific work, then, is to break down this heterogeneous group and find more specifically defined physiological diseases. But how? Which questions should be asked? Depression in one or another form is an exceedingly common condition, much more common than schizophrenia, and rivals it as the Nation's number-one mental health problem. According to an NIMH estimate, 15 percent of all adults between 18 and 74 may suffer significant depressive symptoms in any given year (NIMH 1973).

MANY QUESTIONS AND A FEW ANSWERS

A great deal is known about depression already. But there are still more questions than answers. Depression can be provoked by a variety of life stresses, such as losing a job, losing a mate, getting sick; the causes of a depressive episode are endless. So how should we categorize patients in order to increase the chances of finding specific biological subgroups and, more important, specific therapies for each? Researchers have approached the problem from several directions: Is everyone depressed sometimes? What is a psychotic mood disorder? How do the affective disorders, sometimes divided into unipolar and bipolar depression, differ? Are women different from men in their expressions of depression? Are depressed young people different from depressed older people?

Some of these questions have at least partial answers. True, some of these partial answers have led to more questions, but the search has also led scientists to the point where a major, all-out research project, now under way, should provide an important new understanding of this condition.

Is Everyone Depressed?

Three St. Louis scientists, Woodruff, Clayton, and Guze, say "no." They interviewed 900 relatives of 500 patients and nearly half said they could not remember ever having more than a few days of dysphoria, nor did they report symptoms of mood disorder. However, some 14 percent of the relatives had signs of depression, along with mania to which it is sometimes linked.

"Regardless of our preconceptions, it seems unquestionable that a large number of persons lead relatively unruffled affective lives . . . although depression is common, freedom from memorable dysphoria is even more common. If the 'mass of men lead lives of quiet desperation,' that despair is to be seen as philosophical, economic or existential, not psychiatric in our sense. We believe the distinction is important and that failure to appreciate it has been a constant source of confusion for psychiatrists." (Woodruff, Clayton, and Guze 1975.)

What is Psychotic Mood Disorder?

If any distinction is basic to psychiatric medicine, it is the distinction between psychotic and nonpsychotic disorders. Among its various uses, the word "psychotic" means severe and incapacitating mental illness. "Neurotic," on the other hand, traditionally refers to the less severely affected, still functional patient whose contact with reality is intact and who can be more easily treated with psychotherapy. While definitions and uses of these concepts have changed over the years, the basic distinction still carries clinical weight. Is this distinction important in mood disorders? What does it mean? The St. Louis group also asked that question and the answer was surprising. There really aren't significant differences between psychotic and nonpsychotic depressive patients, except for symptoms usually associated with psychosis: the presence of hallucinations, of delusions, or of ideas of reference (seeing personal meaning in random events, for example, "That stranger changed his seat; he is trying to tell me something."). Psychotic patients had not been more neglected or abused by parents; there was, in fact, no discernible difference in home environment, parental illness, marital history, drug abuse, age, education, or number of suicidal attempts. The only difference was a history of hospitalization—the psychotics had been there before.

Even more surprising was the lack of difference in any symptoms other than those used in the original diagnosis. Psychotics were no different than nonpsychotics on 75 common psychiatric symptoms, including nervousness, anxiety, fatigue, fainting, visual blurring, paralysis, unconsciousness, vomiting, abdominal pain, dysmenorrhea, sexual frigidity, impotence, irritability, temper outbursts, fighting, wanderlust, insomnia, back pain, joint pain, depressed mood, inability to work, trouble thinking, trouble concentrating, hopelessness, lack of interest, guilt, fears, obsessions, compulsions, and so on (Guze, Woodruff, and Clayton 1975).

This lack of difference reflects the difficulty scientists are having in finding useful distinctions or subgroups within the depressive disorders. Two groups of patients may look quite different. One may be agitated, another may sleep all day. In one group, the patients hallucinate while a second group may seem more rational, even though they are emotionally disturbed. Those look like important distinctions from the perspective of clinical behavior. Yet, when one goes beyond the single trait, distinctions wash away. The groups don't actually differ in any other way; the symptoms do not lead to biologically distinct disease processes.

"Classification based on clinical symptoms alone is of little practical value," concluded the Guze study of 108 depressed individuals. "Any differences which existed in this study can be best used as starting points for further analysis."

It is not surprising that this should be the case. Depression is a massive piece of behavior, and, expectably, it should reflect the operation of multiple systems in both the psychological and biological realms. To trace the symptoms of depression to a specific biological-genetic factor is something like tracing muscular aches and pains to a specific virus. Usually it is not possible to diagnose disease from an ache in the body.

How Do the Affective Disorders Differ?

Underlying the affective disorders, unipolar and bipolar depression, are such psychopathological factors as "depressive mood, feelings of guilt and worthlessness, hostility, anxiety-tension, cognitive loss and subjective uncertainty, loss of interest and involvement in activities, somatic complaints, sleep disturbance, motor retardation in speech and behavior, bizarre thoughts and behavior, excitement, and denial of illness" (Katz and Hirschfeld 1978, p. 1186). This observation, gleaned from reports of how patients see themselves and how they are seen by physicians and nurses, provides a rather complete characterization of individuals diagnosed as "depressed."

Over the years, classification of the several clinical types of depression have changed and developed to accommodate new genetic and biological evidence. For instance, the old, reliable concept, "endogenous depression," refers now to a composite of clinical psychopathology which may include waking early, loss of appetite and weight loss, daily mood variations, psychomotor disturbance, little reaction to outside stimuli, and, of course, severely depressed mood, a condition usually responsive to imipramine and chlorpromazine. According to the Collaborative Drug Study on Depression conducted by NIMH's Psychopharmacology Research Branch, patients in this category are, on the whole, older, more severely ill patients who evidence depressive feelings of hopelessness and worthlessness (Katz and Hirschfeld 1978).

As attempts at better classification continued, questions arose as to the distinctions between, for instance, endogenous versus reactive and agitated versus retarded depression. For research purposes, the distinction between *primary* and *second-*

ary depression proves more acceptable to many investigators. A primary affective diagnosis designates a patient whose premorbid history is good, who has no previous mental disturbance to report, or whose previous attacks were mania or depression *only*. The diagnosis of secondary affective disorder is given to an individual experiencing an affective episode who has had a previous psychiatric illness diagnosed as a condition *other than* a primary affective disorder (Robins, et al. 1972).

The most common distinction currently seems to be the unipolar-bipolar one. The unipolar diagnosis, more common by a ratio of 10 to 1, according to one estimate, refers to the classic idea of depression, with its symptoms of withdrawal, lethargy, anxiety, and feelings of grief and guilty self-blame. Bipolar, as its other name, "manic depression," suggests, indicates the occurrence of both a manic and a depressive episode, either separately or chronologically linked. This distinction is especially important because many researchers are beginning to agree that the two, unipolar and bipolar, may be different disorders, possibly with different genetic components, biochemical factors, and family linkages.

Do Men Differ From Women in Depression?

The sexes differ little in depressive illness except for the increased incidence of the illness in women, but there is an important fact that needs to be explained. In families with unipolar symptoms, the extra prevalence of female depression is matched by alcoholism among men, so that, if alcoholism and depression are lumped together, men and women are equally ill. Not everyone thinks the two groups should be considered together, but George Winokur thinks they are related. As a result of his several family studies, he and his colleagues group alcoholism, sociopathy, and depression under the one name, "depression-spectrum disease," and theorize that the same genetic predisposition may manifest itself as depression in women and alcoholism or sociopathy in men (Winokur 1975, p. 16). In fact, they define *pure* depressive disease as occurring in any rigorously diagnosed depressive who does *not* have an alcoholic or antisocial first-degree family member. They hypothesize further that depression-spectrum disease is an illness that affects mainly the female relatives of alcoholic men (VanValkenburg et al. 1977).

In other ways, men and women manifest their disorder quite similarly when they become depressed. In one study of 100 patients, the sexes differed on only 2 of 41 symptoms. The women were more likely to say they couldn't cry, while the men were more likely to have more physical complaints (Baker et al. 1971).

Are Young Patients Different From Older Patients?

The question of age in depression refers, not to a generation gap, but to the age at which an individual experiences the first episode of illness. Increasingly, there appear to be important differences between patient groups who become depressed before the age of 40 and those whose first episode comes after 40. The early group has considerably more affective illness in the immediate family. In the first group, about 20 percent of the relatives show affective illness on the average, while in the second group, whose illness comes later in life, family disorder runs about 10 to 12 percent (Winokur 1975, p. 17). If alcoholism is added to the calculations, illness in the early group rises to about a quarter of relatives. In families of middle-aged patients, however, alcoholism appears infrequently. This difference, plus the sex difference, leads Winokur to believe that early-onset depression belongs to a spectrum of disorders which has a different genetic component than does the late-onset condition.

Others see the age distinction in a different light. William Bunney Jr., of NIMH believes early-onset depressives may actually be unrecognized bipolar patients, while the late-onset group represents pure depressive disease. Whatever the case may be, the people who become depressed early in life are likely to have a fairly high rate of psychiatric trouble, including mood disorder, among their relatives.

Do the two groups also act differently or show distinct symptoms? The answer is "yes, somewhat." The younger group tends to feel more guilt, more tears, more worries, more irritation. Although members of both groups may feel equally suicidal, the younger ones are more likely to make suicide plans and to tell others about them. Finally, the early-onset group feels less mentally agitated and disordered in thought process. By comparison, the late-onset depressives show fewer psychological responses such as guilt. They are also more constipated and more prone to lose weight (Baker et al. 1971). These differences may seem a consequence of maturity, but a test of that thesis suggests not. The other symptoms appear to reflect mental and emotional traits that persist over time.

MANIC-DEPRESSION

The isolation of *bipolar illness* as a distinguishable entity in which mania and depression are linked is one of the most important success stories in the field since depression was first recognized. It suggests that the search for its genetic components, for family linkages, and, indeed, for its behavioral characteristics should be a dramatic account. Such an account, too, should illumine the dark places in our knowledge of other forms of depressive illness.

This is the recurring dream of a man with manic-depressive illness: "I am riding a motorcycle around the rim of a large bowl and I am afraid. If I go too fast, I will fly off the edge into space. If I go too slow, I will fall into the middle and be crushed. I have to pace myself so that I stay on the narrow rim."

Manic-depressive illness is a disturbance of mood, and of rhythm and energy levels as well. Individuals with this condition are in a state of near-constant change, from lethargy to excess activity, from hypersomnia to insomnia, from blunted feelings to overwhelming emotions.

At one extreme lies mania wherein the individual may stay up all night in a prodigious outpouring of feeling and emotion and apparently feel no fatigue the next day. Contrary to popular belief, mania is not usually a pleasant state of mind. The engines of emotional drive are turned up too high, and they easily slip gears from one strong feeling to another, from positive to negative emotions. Hostility, belligerence, even desperation, are as common to the manic state as is talkative sociability. Unfortunately, euphoria is not a very stable, or even frequent, experience.

At the opposite extreme lies depression—which in this condition takes the form of drowsy lethargy and social withdrawal. Unlike most depressive illness, which tends to contain a fair amount of agitation and sleeplessness, this type of depression often leads to excess sleeping. Even when awake, the manic-depressive patient tends to lie around from sheer lack of drive. He may be sad, but more frequently the emotional state is simply blunted and inactive.

The sleep patterns of such a patient dramatically illustrate the cyclical nature and continuous changes of manic-depressive illness. Some 10 patients have been monitored day and night for months at NIMH's sleep laboratories, so that there is now a large body of hour-by-hour data on the sleep cycles, circadian rhythms, biochemistry, and emotion of depressed persons.

One Woman's Story

Fortunately for this research, one patient was euphoric rather than aggressive during manic episodes, a circumstance which made her more cooperative than many patients, who are often too belligerent to cooperate with the recording apparatus, or so restless that they do not easily tolerate the restrictions of electrodes and cannulas. Although her cyclical mood changes were striking, they were probably not incapacitating enough to require inpatient treatment, except for a history of suicide attempts following the blunted withdrawn mood and feelings of social inadequacy which occurred during the periods of depression.

Her depressions typically lasted about 20 days; a week of mania came about every 6 weeks, and the rest of the time she was relatively normal. Her sleep cycles and circadian rhythms marked off the daily changes, as this thumbnail sketch, beginning with the gradual switch from mania to depression, shows:

> The cycle begins with the late manic phase, a time when the patient seems fairly normal. She goes to bed about 11 or 12 at night and is awakened at the routine hospital hour of 7 in the morning. Her temperature peaks in the early afternoon, following the normal circadian rhythm wherein the temperature reaches a low at night and a high at midday.
>
> Then the drowsiness begins. Each night she gets sleepy earlier and earlier in the evening, until she is going to bed at 9:30 or 10 o'clock. Her body temperature, meanwhile, also peaks earlier in the day, around noon, indicating that the daily period of high energy has gradually shifted to an earlier part of the day. At this point, the patient, although appearing to be in a manic mood, doesn't want to do anything in the evening except doze. At home she turns down parties, to the consternation of family and friends who expect her to be more sociable in this phase.
>
> Now depression begins to set in. The circadian rhythms reverse themselves. Instead of getting drowsy earlier and earlier in the evening, she begins staying awake later and later. She goes to sleep about 10 minutes later each night for the next 20 days, shifting in total about 3 hours, from 10 p.m. to 1 a.m. Body temperature makes an even more dramatic shift, peaking later and later until the energy-temperature peak comes at 5 or 6 in the evening. She

probably would sleep until late in the morning if the hospital permitted, but she is awakened at 7 and feels terrible in the morning. The only time she is active and feeling good is during the evening.

So far, all the changes have been gradual, melting one into the other. Now a dramatic shift occurs—the onset of mania. It comes in the middle of the night for no obvious reason. Suddenly the depression is gone and the patient is awake, energetic, talkative, and ready to roar. The next night, she doesn't sleep at all. She runs around the ward seeking out people to talk to, becomes preoccupied with her appearance, and wears a lot of makeup.

In the nights following the manic switch, sleep time begins to lengthen again and the woman wakes up increasingly later in the morning until she is back to normal at 7 a.m. Now, circadian rhythms reverse themselves again and the peak temperature-energy point begins to sink. Drowsy evenings reappear and the cycle begins again.

UNIPOLAR DEPRESSION

Depression alone, without mania, or *unipolar depression,* occurs more often, by a ratio of about 10:1. Even in families with known bipolar illness where mania is obviously present, many of the affected members will show only depression, never mania. It is still not known whether they have the same genetic constitution as their bipolar relatives, but it is known that the clinical signs of the condition differ substantially. Depression usually appears first, years before mania, so that the mean age of onset for depressive episodes is 26, while for manic episodes it is about 30. This was aptly illustrated in a St. Louis study of 61 manic patients, 27 percent of whose first-degree relatives had affective illness. Seventeen percent had depression; another 10 percent had mania as well (Winokur, Clayton, and Reich 1969).

Evidence for the genetic transmission of unipolar depression is less persuasive, and attempts to distinguish various types of this form of the illness have been less successful than in the case of bipolar depression, even though it is statistically the more common of the two. Pure depression, according to some investigations, is seen equally in males and females, and, in terms of rates of illness, there is equality between parents and siblings of the depressed person. Winokur sees both pure depressive disease and depression-spectrum disease as unipolar, the former occurring typically in late-onset male patients and the other in early-onset females, with familial alcoholism a part of the picture. His hypothesis is that a dominant gene, possibly more than one, causes depression (Winokur 1975).

MORE QUESTIONS FROM TWIN AND FAMILY STUDIES

Twin studies have produced evidence for genes in unipolar depression, although not to the satisfaction of a number of researchers. Joseph Mendels found identical twins showing a very high rate of concordance for mood disorder (1974). Based on a review of six twin studies, he found that, in 74 percent of the cases, if one of the pair was sick, the other was likely to be also. The suggestion of the significant role of genes came from the comparison of the rate for MZs with that of fraternal twins, who showed only 19 percent concordance. Another study found stronger evidence, with 12 identical twins reared apart, who continued to show a high concordance rate for affective illness even though they did not share the same household (Price 1968).

Is Heredity a Primary Factor?

The association of manic and depressive moods had been recognized clinically for years by scientists both in this country and in Europe, but the awareness that these might constitute a specific biological illness came only recently. The researchers at Washington University in St. Louis conducted family studies on several hundred depressed patients to see whether distinctions in family background might further identify the heterogeneous depressive groups. They found that mania was overwhelmingly concentrated in patients with a family history of two generations of affective illness. Patients without such a family history had virtually no mania, although they were depressed. It was the only difference that separated the two groups. Mania was a crucial symptom. It indicated the presence of a specific illness that clustered in given families. Moreover, it was accompanied by a fairly high rate of illness among parents, siblings, and offspring of patients.

Reported rates of illness vary from study group to study group. One of the highest came recently from Mendlewicz (1975) who examined 781 family members of 134 bipolar patients receiving lithium therapy at the New York State Psychiatric Institute. Nearly 40 percent of the 781 parents, siblings, and children were considered to have affective illness. This is a very high risk rate compared to other studies reviewed by Gershon (1976), where the risk to first-degree relatives varied from 11 percent to 30 percent. This enormous variation in familial rates for affective illness can be attributed partly to different diagnostic methods and criteria. Criteria for identifying relatives with depression or mania have not been the same from study to study. Also, the patient populations may differ in unknown ways. One group may be sicker than another or more biologically impaired. The New York study, for instance, might have a group that is more genetically loaded for affective illness than are other patient groups, thus accounting for the higher rate of familial illness.

Whatever the case may be, relatives of manic-depressive patients have a much greater chance than normal of developing an affective disorder. The usual risk for manic-depressive disorder in a general population varies from less than 1 percent to 2.5 percent, according to studies done in five European countries (Winokur 1975, p. 9).

These general population rates form a provocative contrast to those of relatives of persons studied in the early sixties by the St. Louis group. Winokur and Pitts (1965) found that among 366 patients with mood disorders, both unipolar and bipolar, the rate of illness among mothers was 22.9 percent. Fathers had much lower rates, at 13.6 percent, reflecting the common opinion that women show about twice as much affective disorder as do men, or, at least, they receive a diagnosis of depression more often than men do. These rates can be compared to the patients of a control population, 180 patients with physical illness, whose rate for affective illness was only 1.7 percent. The difference is striking: 1 percent as against 13 percent for fathers and 22 percent for mothers. Obviously mood disorders run in families, and clearly some families carry a substantial risk.

Does This Mean the Cause of Depression Is Genetic?

Family studies do not in themselves establish genetic proof, because the disorder may be rooted in the family environment as well as in the family genes. Family studies do, however, point the way for further genetic research and have been particularly valuable in the work on mood disorder.

The preponderance of depressive illness among women led the St. Louis researchers to consider the possibility of X-linkage, which refers to a gene carried on the female sex chromosome. These genes tend to sort themselves in a characteristic pattern, and since females inherit two X chromosomes and men only one, an X-linked condition might appear twice as often among women, or women may carry the gene even if they are not ill themselves. Since a father contributes only the Y chromosome to his son, he can give an X-linked trait only to his daughter, who receives his X chromosome. Thus, ill daughters should have equal numbers of ill mothers and fathers, since they, unlike the sons, can get the X chromosomes from either parent.

Could X-Linkage Be a Factor?

The St. Louis team culled 49 parent-child pairs from a cohort of manic-depressive patients. There was the expected distribution: The daughters had equal numbers of ill mothers and fathers, and sons had only ill mothers in 14 of 15 cases. But there was one exception, a single case of father-son transmission. Perhaps it could be explained in some other way. The X-linked hypothesis still looked good, and the researchers pressed on.

The next step was to find a marker gene known to be on the X-chromosome and to associate that with affective illness. Early work attempting to link manic-depressive illness with blood groups had not been very successful, but the St. Louis workers were able to find two families in which bipolar illness was strongly associated with color blindness, a known X-linked gene.

In one family, five people had affective disorder and all of them were either color blind or carriers of the color-blind gene. In the second family six people had affective disorder, and they also were color blind or carriers of the gene. A search of the family tree in one family turned up more than 150 relatives among whom color blindness or a carrier state and emotional disturbance coexisted in all but one case (Winokur, Clayton and Reich 1969, p. 123).

It was a dramatic finding. For the first time someone had found specific evidence for genetic transmission in the mood disorders. From now on, genetics would be in the forefront of research on depression. In the same 1969 publication the St. Louis researchers stated that "the finding of an X-borne gene . . . proves a genetic factor in manic-depressive disease" (p. 41).

That statement may have been premature. Subsequent research has turned up a considerable amount of father-son illness which would not occur with an X-linked trait. These results have led some, but not all, researchers to conclude that the preponderance of mood disorder among women is probably due to environmental causes, or to some genetic factor other than transmission of the X chromosome.

The color-blindness studies, the other approach to X-linkage, met a different fate. Subsequent work by Mendlewicz and Fleiss (1974) in New York strongly supported the St. Louis findings. In 17 families they found a close association between bipolar illness and two types of red-green color blindness, called "protanopia" (red deficiency) and "deuteranopia" (green deficiency). They also found an association, but less close, between bipolar illness and the X_g blood group, another X-linked trait.

Still, there are methodological problems with the linkage hypothesis. To clear up some of the confusion, two NIMH scientists, Drs. William Bunney and Elliot Gershon, joined with Dr. Winokur to evaluate the state of the art. The three had this to say (1976): "Linkage appears to be present . . . the data are compatible with the hypothesis of X-linkage. . . . But in view of statistical problems, it would be premature to conclude that linkage is an established finding."

In that same review, the authors supported the existence of X-linked affective illness for at least some people with manic-depressive disease. The evidence is strong enough to indicate "that there may be some subpopulations of manic-depressive patients in which the illness is transmitted on the X chromosomes." Winokur believes that 51 percent of manic-depressive patients have the X-linked condition. Others are reluctant to place the figure so high. And so the search for confirming evidence goes on.

CONTRIBUTION OF FAMILY STUDIES

Family studies are being used to show correlations between mental illnesses of various sorts and sex, age, and period of onset; family members of the same and different generations; and physical and socioeconomic conditions. They are proving valuable, too, in probing possible connections between unipolar and bipolar depression, depression and schizophrenia, and depression and sociopathy.

A large, ongoing study is the "Iowa 500," so named because it is a combined followup and family study of around 500 schizophrenic and primary-affective-disorder patients who were discharged from the Iowa Psychopathic Hospital about 35 years ago. It all began in 1971, when Winokur, newly arrived from

Washington University to join the faculty of the Department of Psychiatry at the University of Iowa College of Medicine, discovered "a series of amazingly comprehensive medical charts" in the basement of the hospital. The charts covered hospital admissions in the 1930s and 1940s, complete with verbatim transcripts of patient interviews, staff meetings, and minute examination of history and mental status of the patients.

After selecting from these documents patients clearly fitting the diagnosis of schizophrenia or primary affective disorder, Winokur, Ming Tsuang, and their colleagues began the search for data covering the intervening years. They hope, by this means, to learn much more about the symptoms, course, and outcome of the disorders. Equally important is the opportunity for the large family study of siblings and children of the original cases. Since these relatives, 35 years later, have nearly all passed through the risk periods for the two disorders, measurement of family morbidity risks can be accurate, rather than estimated (Tsuang and Winokur 1977).

The specific research criteria classified 200 cases of schizophrenia, 100 of mania, and 225 of depression. From the Department of Surgery roster of admissions from the years 1938–1948, 160 persons were selected to serve as controls and sampled randomly as to sex and public or private admission status to provide a representation proportionate with that of the index cases. Working blind as to diagnosis of their subjects, interviewers obtained information on current address and living situation, marital status, employment, and psychiatric status. From the spring of 1974 to June 1, 1976, 661 (97 percent) of the 685 index cases had been followed up, and preliminary conclusions reported (Tsuang 1976).

Preliminary results of the family-study portion of the "Iowa 500" have been applied to the problem of differentiating and subtyping schizophrenia and affective disorder. According to Tsuang, "Schizophrenia, affective disorder and control relatives were clearly distinguished according to their rates of schizophrenia and affective disorder, respectively: schizophrenia relatives having a significantly higher rate of schizophrenia than the affective disorder and control relatives, whose rates were not significantly different; affective disorder relatives had a significantly higher rate of affective disorder than either the schizophrenia or control relatives, and the rates for the latter groups were not significantly different from each other. . . . Among the affective disorder relatives we found no significant tendency for subtype to be associated with like subtype in the index cases, that is, for unipolar to go with unipolar, bipolar with bipolar." The researchers attribute the latter to methodological problems (1977, pp. 10-11).

Not all of the results are in from this wide-ranging study. Findings must be analyzed further, after even more complete information on both living and deceased relatives and index cases has been gathered and subjected to assessment techniques, providing more substantial foundation stones to the pathway.

THE COLLABORATIVE STUDY

The NIMH Clinical Research Branch Collaborative Study of the Depressive Disorders has subjected these accumulating biological and genetic leads to even more intense scrutiny and evaluation. Nine major centers have joined in this effort which will bring to bear half a dozen major scientific approaches on one group of patients.

The first part, the biological-studies component, is the largest study of its kind, with patients drawn from five centers and combined into a single group which can then be studied from several vantage points—pharmacological, hormonal, biochemical, genetic, and clinical. Its aim is to isolate one or more specific disease processes that affect mood disorder. The hope is that this project will succeed where others could not be sufficiently definitive.

Some 20 years of biological research on depression have yielded at least three major theories of the disorder, in addition to a multitude of separate biochemical findings. Scientists expect that if they can get together a large enough group of patients to be examined and treated by the combined expertise of five research centers, they will be able to find proof for one or more of the biological theories. The difficulty in the past has been that no one center could study enough patients in the necessary detail. The biological-studies component of the collaborative project will ultimately have nearly 200 patients and 141 controls, all selected by the same criteria. The five centers are located hundreds of miles apart—Philadelphia (the Veterans' Administration Hospital); St. Louis (Washington University); New Haven (Yale University School of Medicine); New York (Cornell University); and Chicago (University of Chicago), so the project must be kept together via the airlines. For example, once every 2 months, technicians from each center take a freezer full of blood and urine samples to a central location where they meet and exchange bottles: hormonal samples to Cornell, electrolyte samples to Philadelphia, drug samples to St. Louis, and so on. In addition, each patient is videotaped once a week; the tapes are mixed and sent randomly to the various centers for blind scoring of clinical status. In this way, the researchers hope to correlate biological with psychological findings and to bring order to the current jumble of information.

In late 1977 the second phase of the collaborative effort was begun, adding to the study of the biological underpinnings of depression. The second phase is an attempt to increase understanding of personality and of hereditary or familial and other environmental and social factors which may contribute to the development of mania and depression. The collaborating institutions for this phase are Harvard Medical School in Boston; Rush Medical School, Chicago; University of Iowa Medical School, Iowa City; Columbia University College of Physicans and Surgeons, New York City; and, as in the first phase, the Washington University School of Medicine, St. Louis.

HOPES, AIMS, AND A FEW REALIZATIONS

NIMH and the researchers involved hope to integrate the often-competing hypotheses about the origins of depression by developing new methods that can be applied in a standardized way from institution to institution. To begin with, the scientists realize that they must find reliable and standardized methods for grouping patients, a basic necessity because of the heterogeneity of the cause, symptoms, and responses to treatment of the affective disorders. A schedule of diagnostic criteria has already been produced by a pilot-study group for the initial selection of patients for both the biological and clinical phases of the project. It is hoped that, from the many systems now in use for grouping patients, the system to be evolved will be useful both for scientists studying the disorder and for clinicians who treat it. Other goals are to arrive at a system which can predict both course and outcome of the illness, can be used for patients from any social or economic background, and can tell physicians which classification system can be used most effectively in the selection of treatments. The clinical programs are being conducted in much the same way as are those of the first collaborative study, with patients tested and treated in all five centers, after which the data will be analyzed at the individual centers and synthesized centrally.

Throughout, an effort has been made to distinguish pure depression from all other mental conditions that may have depressive symptoms as secondary components. Depressive symptoms accompany most psychiatric difficulties, such as schizophrenia, hysteria, compulsiveness, drug abuse, alcoholism, and other clinical states. By definition people so diagnosed do not have depressive illness and will be excluded from the collaborative study, which is using the relatively new clinical categories of "secondary" depression and "primary" depression. The primary group is as "pure" as researchers can currently make it, that is, the patients show only mania and depression or depression alone, without any other psychiatric symptoms. This group is thought to be genetically predisposed to affective illness.

SOME SOLID SIGNS

The scientists participating in the collaborative studies are working at several levels along the pathway from genes to behavior. There are several levels of physical functioning, from the most basic function of the nerve cell to the action of circulating hormones in the body, each level contributing in its own way to the pathway.

Electrical Activity

The most basic level involves the electrical activity of the nervous sytem, the most primitive aspect of nerve function in an evolutionary sense. Nerve function is carried out by the action of electrolytes, the medium through which current is passed. According to the membrane theory of nerve conduction, electrically charged ions (called "cations") of sodium, potassium, calcium, and magnesium, among others, penetrate back and forth across the nerve membrane, setting up the disequilibriums required for the cell-firing function. With each stage of activity, the cations change their concentration on each side of the membrane, and it is the distribution of these cations that determines the "wave" or transfer of impulse along the nerve axon. Even the most primitive nerve function has these electrolytes, for without them, nerve function does not exist.

Researchers believe that some depression may be traced to a malfunction in the cation system, either in the distribution of cations across the membrane or in the membrane itself. Several studies have reportedly found a characteristic distribution of sodium and calcium in groups of depressed patients. These reports, although provocative, are unproven and sometimes in conflict with each other. To confirm just one association between depression and electrolyte distribution would be an important advance.

A key to such work is *lithium,* the now common treatment for bipolar depression. Lithium, which bears a molecular structure similar to sodium and calcium, functions like a cation in the nerve cell, so it can be used as a model for what is happening in these biological pathways. Recently, a Philadelphia team led by Mendels found that the red blood cells (RBC) of bipolar patients took up and concentrated lithium at higher levels than did blood cells of other depressed patients. These scientists believe the higher blood-cell concentrations reflect a difference in cell-membrane characteristics which also obtains in brain cells.

Thirteen patients were tested in the study, and eight, all but one of whom was diagnosed as having bipolar illness, responded to lithium carbonate with an improvement in mood. The eight had higher lithium concentrations in blood cells than did the five who did not respond to this drug therapy (Mendels and Frazer 1973).

Thus, the current state of the art suggests that the genes could be engendering bipolar illness through changes in the nerve membrane, creating a maldistribution of cations, thus causing significant change in nerve-cell activity. Suppose that is one genetic pathway for a type of affective illness. How many genes might be involved? Some investigators believe a single dominant gene could cause biological vulnerability to manic-depressive disease. A more likely case is that many genes are implicated, although only one may have major effects. The genetic base for electrolyte distribution is complex, as described by Mendels:

"There are three to five cations, each balanced in amounts relative to others. Just take one part of that, the amount of sodium inside the cell relative to the outside. That doesn't have one genetic regulator. It probably has five or six regulatory systems playing different roles: two or three to determine how much sodium moves in and two or three determining how much sodium moves out. Then you get into all the interactions between sodium and other cations."

Neurotransmitter Activity

The second level of biological function comes a step closer to behavior. One of the functions of cations is to make possible synthesis, storage, and release of chemical messengers at the nerve endings (synapses), which either activate or inhibit other nerves that impinge on that ending.

These messengers, or neurotransmitters, form a crucial link between the individual neuron and the rest of the brain. They modulate activity in the brain structures that subserve behavior, and they have for years been implicated in the mood disorders. They can have an inhibiting or exciting effect on whole brain structures that underlie emotional behavior. They link electrolyte action with hormonal action, a third physical level implicated in depressive illness. Finally, they can affect the distribution of cations, as well as be affected by them. A sodium imbalance in the cell, for instance, may cause a serotonin deficiency, which leads to poor sleep. But the system can also work in reverse. A serotonin deficiency might disrupt the sodium balance, which then leads to further trouble. This hypothetical example, not proved in any of its details, illustrates an important fact: The biological pathway is clearly there for genes to affect major emotional behaviors.

New Clues

Some of the neurotransmitter clues that will be tested by the collaborative project include the following:

1. Spinal taps show that serotonin levels might be low in some depressed patients, particularly psychotic individuals. Interestingly enough, it may also be low in some schizophrenic individuals. It is quite possible that a serotonin deficiency could reflect a general psychotic state rather than a depressive condition alone.

2. One of the more recent transmitter theories suggests that mood state is determined by a balance between two major chemical systems, one of whose primary functions is to inhibit transmission and that of the other to excite transmission. To give empirical substance to this theory, scientists will have to be able to localize studies in given regions of the brain. Measurements of whole-brain chemistry tend to produce little relevant information.

3. Work is focusing on the emotional centers of the brain where depleted inhibitors could be allowing excess neural activity. One crucial location is the hypothalamus, a versatile brain structure which is a locus for triggering intense emotional responses and which affects the regulation of sex, appetite, sleep, and aggression. It also exerts control on the pituitary gland, which in turn controls hormonal releases throughout the

body. Thus, the hypothalamus is an important gateway from brain to body via the pituitary. It is here, in the hypothalamic-pituitary axis, that some researchers hope to find the key to affective disorder.

Many of the symptoms of affective disorder can be explained by a malfunction of transmitters in the hypothalamus. The locus would explain the poor regulation of circadian rhythms and the changes in sleep, appetite, and sex. It would account for disturbances in motivation and arousal, because the hypothalamus is a primary center for integrating motivated behavior. It is the place where emotional drive meets up with the once-instinctive behaviors of sexual activity and eating. What this suggests is that stimulation in brain structures below the hypothalamus, the more ancient areas sometimes called "the old brain," can elicit fragments of behavior that have no emotional meaning or coordination. Stimulation in the hypothalamus, by contrast, brings forth integrated behavior, complete with drive, purpose, emotional expression, and learning (Quarton, Melnechuk, and Schmitt 1967, p. 565). Moreover, the pathways that provide pleasurable and aversive stimulation both run through the hypothalamus, so that lesions in different parts of this brain region can produce strong negative or positive reinforcement.

Finally, the hypothalamus controls many functions of the autonomic nervous system of the body. It tells the pituitary gland when to release hormones regulating sexual cycles and stress reactions; it provides the means by which emotional state affects the physiological response. The so-called "vegetative" symptoms of depression, which include such things as a change in gastrointestinal motility can be explained by a lesion in the hypothalamus. The evidence for the possible existence of such a malfunction may come from yet another physiological explanation of depression.

The neurotransmitter theory, also called the "catecholamine" or "amine" hypothesis, once dominated thinking on the biology of depression. By this theory, depression resulted from a depletion of certain transmitters in the brain, especially norepinephrine and serotonin, which were associated respectively with arousal and sleep. Mania, on the other hand, was thought to be the result of an excess of these neurotransmitters. In fact, the neurotransmitter theory derived originally from discoveries of how mood drugs act on the brain. Drugs which deplete neurotransmitters usually cause depressed behavior, while drugs that raise levels frequently cause mania. But the neurotransmitter theory is having to undergo some modifications. Efforts to reverse depressive symptoms by giving patients the chemical precursors of neurotransmitters have not for the most part been successful.

Most notable is the failure of tryptophan to have much effect on depression and sleep. Tryptophan, an amino acid found in the average diet, changes to serotonin in the brain, so that if a deficiency of serotonin is a cause of depression, a still valid hypothesis, then logically tryptophan should relieve depression. But the evidence is meager that it does. Nevertheless, trytophan is becoming a popular item in health-food stores, where it is sold in pill form. Clients of the health-food subculture may have heard it is either a mood elevator or a sedative.

Despite this negative evidence, there is still reason to believe that neurotransmitters play a crucial role in mood disorder. Findings of the collaborative team, corroborating or disproving present hypotheses and adding new ones, should do much to make the pathway clearer.

Endocrine Activity

The third physiological explanation of depression, the "endocrine" theory, may help explain, among other puzzles, those "vegetative" symptoms of depression, equivalent in effect to a lesion in the hypothalamus. Depressed patients show disturbances in several hormonal systems, including sex, thyroid, and adrenal-gland hormones. Attention is focused on one of these, cortisol, which is released by the adrenal gland to prepare the body to cope with stress. Cortisol has its main impact on metabolism, causing, for instance, salt retention and sugar elevation, both stress reactions. It also affects mood and can bring on either mania or depression, depending on the individual and his state of mind.

Recent evidence collected on this critical hormone now shows that 40 to 50 percent of depressed patients have abnormal cortisol secretion. Not only do they secrete too much cortisol, their daily rhythm of cortisol activity is abnormal. The normal pattern of cortisol release in someone who sleeps from 12 midnight to 8 a.m. is to have several large bursts between 5 and 10 in the morning. From this morning peak, cortisol levels off to moderate secretion during the day, falling to nothing in the evening around 8 p.m. and then continuing at low levels during the night.

In depressives, it looks very much as though the system does not shut down. After the morning bursts, depressives continue to secrete cortisol in the afternoon and evening, as though their bodies were being continually prepared for stress. This secretion lasts even during sleep, and, in fact, the greatest difference between normals and depressed patients is in the nighttime hours between late evening and 4 a.m., when the latter group continues to pump out cortisol. Nor is the difference due to sleep deprivation. Columbia University researchers deprived normal subjects of sleep for 8 days, and their cortisol secretion did not change (Sachar 1975). Heart patients awaiting surgery have also been tested for cortisol, and their values were normal except for an unusual burst of cortisol at 11 p.m. It turned out that was the hour the orderly came around to give an enema.

To what extent is this cortisol secretion in depressed patients a result of agitation? Surprisingly enough, there is no apparent relationship between anxiety and cortisol secretion. Sachar has not found that anxious and upset patients secrete more cortisol than others. "You also see the hypersecretion in the patients who are not tremendously agitated. When you ask them if they are upset or frightened, they say, 'No, I just feel terrible.'" Furthermore, anti-anxiety medication does not affect the hypersecreting pattern, and schizophrenics, who may also be upset, do not have the cortisol elevation.

Said Sachar: "It looks like a fundamental endocrine disturbance, a consequence of hypothalamic dysfunction."

If the hypothalamus is hyperactive, one result would be an extra boost to the pituitary and hypersecretion of cortisol. The most recent studies of transmitter levels in spinal fluid indicate precisely that—deregulation of the hypothalamus due to a depletion of norepinephrine.

But what is all that cortisol doing to the body and behavior? One possibility is that the metabolic changes could affect disease processes such as diabetes and heart disease. The extra cortisol could cause hyperglycemia in a diabetic-prone individual or high blood pressure in someone with a tendency for salt retention. As for behavior, there is evidence to show that cortisol does affect mood. Individuals with diencephalic Cushing's syndrome hypersecrete cortisol and frequently show mood changes from euphoria to depression. Administered cortisol is also able to cause mood change, and cortisone, a synthetic form of cortisol, is known to have important emotional effects. That the late President John F. Kennedy was treated with cortisone caused some medical concern about the possible effects of this drug on his emotional state.

An important question thus concerns the role of cortisol in depressive illness. Is it purely a *consequence* of a brain dysfunction, or does it also *contribute* to the depressive process?

Most authorities now give cortisol a role in the disease process, at least theoretically. The hormone circulates back on the brain, and there is a possibility that the increased cortisol might be altering membrane function and electrolyte distribution (Mendels and Frazer 1973). By the same token, steroid hormones can alter the synthesis of neurotransmitters (Mendels, Stern, and Frazer 1976). This kind of interplay has yet to be proven for affective disorder, but the existence of a feedback loop that may keep the disease process going seems quite plausible.

There is yet one more part to the cortisol story, however, and, in many ways, it is the most interesting. Cortisol secretion requires the prior action of a biochemical that has gripped the attention of many behavioral geneticists today. That chemical is a hormone of the pituitary gland, and it functions to stimulate the adrenals into putting out cortisol. Its name is ACTH (adrenocorticotropic hormone).

According to research reported in our chapter, "Potential Pathways in Man," ACTH may be a biochemical source of arousal, or even of fear, although its precise impact on emotions and behavior remains controversial. Current evidence on affective disorder points to an increase in ACTH as the cause of elevated cortisol secretion, a conclusion drawn from the action of the synthetic steroid, dexamethasone. This is a powerful drug which substitutes for cortisol and in normal subjects suppresses any hormonal release into the bloodstream for 24 hours. It acts on the pituitary to suppress ACTH and thereby shuts down adrenal secretion.

But in depressed patients, the cortisol secretion shuts down for only 2 or 3 hours. Instead of undergoing a 24-hour suppression, depressives bounce back rapidly and begin putting out more cortisol, to add to the synthetic steroid already in the bloodstream. This effect implies that the ACTH mechanism, which normally should shut down when bloodstream levels reach high proportions, continues to operate in many depressed individuals (Carroll and Mendels 1976).

If ACTH is elevated in depression, if it has the emotional, mental, and physiological effects attributed to it by recent experiments on animals and humans, then a major piece of the puzzle concerning affective illness may have been found. But for purists in behavioral genetics, there is an even more interesting thought which goes a little further. The entire ACTH molecule is not needed in order to elicit behavioral effects. A small fragment will do, specifically the first seven to ten amino acids of the long ACTH chain.

That peptide is the same fragment that appears as a hormone in amphibians to change skin color in response to stress. Melanocyte-stimulating hormone (MSH) has the same 10 amino-acid sequence as ACTH, and apparently both have the same effects on behavior. When scientists determine what the skin changes of a frog have to do with the mood changes of affective illness, they may discover some fundamental new things about the evolution of behavioral genes.

Summary

Recent years have witnessed many additions to knowledge about the affective disorders, beginning with their separation into two groups, unipolar and bipolar, on the basis of genetic methodology. Going further, some investigators see two diseases in the unipolar category, grounding their hypothesis on differences in rates of illness in family members. The first of these, pure depression, is found equally in males and females, and in parents and siblings, in terms of rates of illness. There is little alcoholism connected with pure depression. Depression-spectrum disease, the second disease in the unipolar group, however, shows up as depression in females and alcoholism in males, according to some investigators. In addition, some research indicates an X-linkage in the bipolar group and possible linkage with color-blindness on the X chromosome.

The biological and clinical groups of the NIMH Clinical Research Branch Collaborative Study of the Depressive Disorders will consider these hypotheses and many more—the benefits of lithium in the treatment of bipolar illness, the role of the neurotransmitters, family transmission in the affective illnesses, and endocrine functions—in their massive research and clinical program. Five centers for each of these study groups will furnish patients diagnosed as suffering from some form of depression. Each center will bring to bear its scientists and its research and clinical facilities on testing and treatment of individuals and on the collaborative evolution of standard classification and groupings from among the heterogeneous welter of symptoms, the correlation of biological with psychological findings, and the prediction of course and outcome of the disorder, so that physicians anywhere will be able to choose the most effective treatment according to the individual patient's classification.

REFERENCES

Asberg, M.; Thoren, P.; and Traskman, L. Serotonin depression: A biochemical subgroup within the affective disorders? *Science,* 191:478-480, 1976.
Baker, M.; Dorzab, J.; Winokur, G.; and Cadoret, R.J. Depressive disease: Classifications and clinical characteristics. *Comprehensive Psychiatry,* 12(4):354-366, 1971.
Bunney, W.E.; Gershon, E.; and Winokur, G. X-linkage in manic-depressive psychosis. *Neurosciences Research Progress Bulletin,* 14(1):46-55, 1976.
Bunney, W.E.; Murphy, D.L.; Goodwin, F.K., et al. The switch process in manic-depressive illness: A systematic study of sequential behavioral changes. *Archives of General Psychiatry,* 27(3):293-303, 1972.
Cadoret, R.J.; Winokur, G.; Dorzab, J., et al. Depressive disease: Life events and onset of illness. *Archives of General Psychiatry,* 26:133-136, 1972.
Carroll, B., and Mendels, J. Neuroendocrine regulation in affective disorders. In: Sachar, E., ed. *Hormones, Behavior and Psychopathology.* New York: Raven Press, 1976.
Dorus, E.; Pandey, G.N.; Frazer, A.; and Mendels, J. Genetic determinant of lithium ion distribution. *Archives of General Psychiatry,* 31(4):463-465, 1974.
Dorzab, J.; Baker, M.; Cadoret, R.J., et al. Depressive disease: Familial psychiatric illness. *American Journal of Psychiatry,* 127(9):1128-1133, 1971.
Frazer, A. Biological aspects of mania and depression. In: Frazer, A., and Winokur, G., eds. *The Biological Basis of Psychiatric Disorders.* New York: Spectrum Publications, 1976.
Gershon, E.L.; Bunney, W.E.; Leckman, J.F.; Eerdewegh, M.; and DeBauche, B.A. The inheritance of affective disorders: A review of data and of hypotheses. *Behavior Genetics,* 16(3):227 261, 1976.
Guze, S.; Woodruff, R.; and Clayton, P. The significance of psychotic affective disorders. *Archives of General Psychiatry,* 32:1147, 1975.
Katz, M.M., and Hirschfeld, R.M.A. Phenomenology and classification of depression, In: Lipton, M.A.; DiMascio, A.; and Killam, K.F., eds. *Psychopharmacology: A Generation of Progress.* New York: Raven Press, 1978.
Marten, S.A.; Cadoret, R.J.; Winokur, G.; and Ora, E. Unipolar depression: A family history study. *Biological Psychiatry,* 4(3):205, 1971.
Mendels, J., and Frazer, A. Intracellular lithium concentration and clinical response: Towards a membrane theory of depression. *Journal of Psychiatric Research,* 10(9), 1973.
Mendels, J. Biological aspects of affective illness. In: Arieti, S., and Brody, E.B., eds. *Adult Clinical Psychiatry, Vol. III of American*

Handbook of Psychiatry, 2nd ed. New York: Basic Books, 1974, pp. 491-523.

Mendels, J., and Frazer, A. Brain biogenic amine depletion and mood. *Archives of General Psychiatry,* 30:447-451, 1974.

Mendels, J., and Frazer, A. Reduced central serotonergic activity in mania. *British Journal of Psychiatry,* 126:241-247, 1975.

Mendels, J.; Stern, S.; and Frazer, A. Biochemistry of depression. *Diseases of the Nervous System, 37(3):3-9, 1976.*

Mendlewicz, J., and Fleiss, J.L. Linkage studies with X-chromosome markers in bipolar and unipolar illness. *Biological Psychiatry,* 9(3):261-293, 1974.

Mendlewicz, J., and Rainer, J. Morbidity risk and genetic transmission in manic-depressive illness. *American Journal of Human Genetics,* 26:692-700, 1974.

National Institute of Mental Health. *Special Report, 1973: The Depressive Disorders.* Washington, D.C.: U.S. Government Printing Office, 1973.

Price, J. The genetics of depressive disorder. In: Coppen, A., and Ashford, H., eds. *Recent Developments in Affective Disorders: A Symposium. British Journal of Psychiatry,* Special Publication, Number 2, 1968.

Quarton, G.; Melnechuk, T.; and Schmitt, F. *The Neurosciences: A Study Program.* New York: Rockefeller University Press, 1967.

Robins, E.; Nunoz, R.A.; Martin, S.; and Gentry, K.A. Primary and secondary affective disorders. In: Zubin, J., and Freyhan, F.A., eds. *Disorders of Mood.* Baltimore: The Johns Hopkins University Press, 1972.

Sachar, E.J. Twenty-four-hour cortisol secretory patterns in depressed and manic patients. *Progress in Brain Research,* 42:81-91, 1975.

Schless, A.P.; Frazer, A.; Mendels, J.; Pandey, G.N.; and Theorides, V.J. Genetic determinant of lithium ion metabolism. *Archives of General Psychiatry,* 32(3):337-340, 1975.

Schlesser, M.A.; Winokur, G.; and Sherman, B.M. Genetic subtypes of unipolar primary depressive illness distinguished by hypothalamic-pituitary-adrenal axis activity. *The Lancet,* No. 8119: 739-741, 1979.

Tsuang, M.T. "A 35-Year Follow-Up of Schizophrenia, Mania, and Depression: An Analysis of Long-Term Outcome by Marital, Employment, Institutionalization and Psychiatric Status." Paper presented at annual meeting of the Society for Life History Research in Psychopathology, Fort Worth, Texas, October 6, 1976.

Tsuang, M.T. "Familial Subtyping of Schizophrenia and Affective Disorders." Paper presented at annual meeting of the American Psychopathological Association, New York, New York, March 4, 1977.

Tsuang, M.T.; and Winokur, G. A combined thirty-five year follow-up and family study of schizophrenia and primary affective disorders: Sample selection, methodology of field follow-up, and preliminary mortality rates. In: Strauss, J.S.; Babigian, H.M.; and Roff, M., eds. *The Origins and Course of Psychopathology.* New York: Plenum Press, 1977.

Van Valkenburg, C.; Lowry, M.; Winokur, G.; and Cadoret, R. Depression spectrum disease *versus* pure depressive disease: Clinical, personality, and course differences. *The Journal of Nervous and Mental Disease,* 165(5):341-347, 1977.

Williams, J.; Katz, M.; and Shields, J. *Recent Advances in the Psychobiology of the Depressive Illnesses.* DHEW Publication (HSM) 70-9053, 1972.

Winokur, G. Depressive spectrum disease: Description and family study. *Comprehensive Psychiatry,* 13(1):3-8, 1972.

Winokur, G. Genetic aspects of depression. In: Scott, J.P., and Senay, E.C., eds. *Separation and Depression: Clinical and Research Aspects.* Washington, D.C. American Association for the Advancement of Science, 1973. Publication Number 94.

Winokur, G. Heredity in the affective disorders. In: Anthony, E., and Benedek, T., eds. *Depression and Human Existence.* Boston: Little, Brown, 1975.

Winokur, G. Secondary depression is alive and well and . . . . *Diseases of the Nervous System,* 94-99, February, 1972.

Winokur, G.; Behar, D.; Van Valkenburg, C.; and Lowry, M. Is a familial definition of depression both feasible and valid? *The Journal of Nervous and Mental Disease,* 166(11):764-768, 1978.

Winokur, G.; Clayton, P.; and Reich, T. *Manic Depressive Illness.* St. Louis: C.V. Mosby, 1969.

Winokur, G., and Pitts, F. Affective disorder: VI: A family history study of prevalences, sex differences and possible genetic factors. *Journal of Psychiatric Research,* 3:113, 1965.

Woodruff, R.; Clayton, P.; and Guze, S. Is everyone depressed? *American Journal of Psychiatry,* 132:6, 1975.

Chapter VII. Children at Risk

Upon reaching adulthood, approximately 1 in 10 children born to a schizophrenic parent will have the same condition. That is the risk to offspring of schizophrenic individuals—10 to 15 percent. For children of two schizophrenic parents, the risk is even higher—35 to 40 percent (Erlenmeyer-Kimling 1975).

Scientists once believed this risk was primarily psychological, that children became vulnerable through the trauma of living with a schizophrenic parent, with all the communicative and emotional hazards that implies. Years of work have been invested in hope of isolating the specific interaction patterns between parents and children that later put the offspring at risk. No longer is the search limited to psychological factors. Scientists spend increasing amounts of time measuring various biological functions in children at risk in the belief that, if genes play a role in the genesis of schizophrenia, there should be some evidence of abnormal physiology in the young child before he becomes ill.

RISK RESEARCH: A NEW CHALLENGE

The relatively new field, called "risk" research, is among the most demanding in behavioral genetics. Not only must scientists find physiological malfunctions in apparently normal young children, but they must follow the lives of those children for years to find out who actually becomes sick and who does not. Any mistakes along the way, such as choosing the wrong variables to follow, mean that years of research may go down the drain.

Moreover, normal psychophysiology in the growing child is not very well known, so that it is often difficult to judge whether some function in the offspring is normal or excessive. Since the children do not appear sick, they don't differ in obvious ways from other children and the physical findings are subtle indeed.

Considering the obstacles, it is surprising that any differences at all have surfaced between the children of schizophrenic parents and others. Yet here and there distinctions are beginning to emerge. In one recent study, the distinction is some peculiarity of the autonomic nervous system; in another, it is a neurological sign; in a third, the difference turns out to be some subtle deficit in the child's ability to focus attention.

In each of these areas, separate research groups have been able to make preliminary distinctions between the children at risk and the offspring of normal parents or other psychiatric patients. Many of the measures are difficult to interpret: What does an electrodermal skin reaction mean in psychological terms, for instance? How important is an attentional deficit in predisposing toward schizophrenia? Yet, taken together, the findings suggest that some children born to schizophrenic parents are biologically vulnerable. Whether that vulnerability is genetic or not has yet to be tested. Whether it will lead to schizophrenia or not remains unknown, since, for the most part, the children are not old enough to have psychological breakdowns. These forms of psychosis usually appear only after puberty, and schizophrenia is typically diagnosed when patients reach young adulthood.

Too Early To Tell?

What happens to those individuals before the age of 20? What are they like at the ages of 1, 5, 10, or 15? No one knows for sure, but some scientists are making predictions like those in the following study, which is based on a group of Minnesota children (Hanson, Gottesman, and Heston 1976).

The children were only 4 when they were first tested with a battery of neurological, mental, and behavior tests as part of a large child development study run by the National Institute of Neurological and Communicative Disorders and Stroke. The original work had nothing to do with schizophrenia but, because the sample was so large, it incidentally included many parents who had or would have a history of mental disorder. The Minnesota researchers, headed by Drs. Irving Gottesman and Leonard Heston, selected 29 parents diagnosed as schizophrenic and pulled out the records on their 33 children. Unlike others in risk research, this team did not directly test the children themselves for any physical abnormality. The testing had already been done years before by the NINCDS Collaborative Project personnel. Instead, the team predicted that a constellation of biological factors would appear among the 33 children selected as at risk for schizophrenia. These vulnerability factors not only would appear in the risk group; it was predicted they would concentrate in a small percentage of these children, 10 to 15 percent, to correspond with the number of offspring of schizophrenic parents who actually break down as adults.

It was a daring prediction: that a subset of children among the 33 would have (1) poor motor coordination, (2) intellectual inconsistencies such as large IQ changes over time or unexplained gaps in performance from one test to another, and (3) schizoid behavior, including emotional flatness and withdrawal mixed with irritability and instability. Moreover, the three vulnerability factors, or indicators, had to coexist in the same child.

The Vulnerable Ones

When all the data were analyzed, 5 children among the 116 in four risk and control groups had all three indicators. All five of the children had a schizophrenic parent. Many other children in the study had one factor alone. It is not unusual for a child to show intellectual inconsistencies or lack of coordination; lags in maturation do appear among many children who become completely normal adults. What is rare is for all three vulnerability factors to coexist in the same child. The authors of the study believe this may constitute a biological vulnerability to adult schizophrenia. Time will tell. Some of the children are teenagers now and approaching the age usual for positive diagnosis of a schizophrenic break.

If the Minnesota predictions are right, they reveal several important traits about some forms of schizophrenia:

(1) The behavior shown by the five children resembles diffuse brain damage. According to Dr. Daniel Hanson of McMaster University, who did the work for a doctoral dissertation, children with encephalitis show similar types of behavior. That is, they have a variety of symptoms, and almost every part of behavior—mental, emotional, and motor—is affected, suggesting that many areas of the brain may be slightly damaged.

(2) Known birth complications cannot explain these results, since the 5 children were on the whole normal at birth and prenatal care for all 116 children was good. But prenatal factors, such as viral infection or chemical insult to the fetus, might have caused brain damage. The vulnerability could also have been caused by abnormal genes, but there is no evidence to prove a genetic cause.

(3) Growing up with a schizophrenic parent may contribute to the disease risk, but that alone could not account for the biological findings. All but 1 of the 33 children grew up with a schizophrenic parent; only 5 have the three indicators, nor were those 5 necessarily from the families with the most severe illness. Moreover, maternal rearing was not a critical factor since, in three of the five cases, the schizophrenic parent was the father, not the mother.

An important feature of this study is that, exclusive of the five children having three indicators, researchers found no significant differences on these measures between the normal group and the risk group taken as a whole. That is to be expected since the rest of the children at risk would actually have no predisposition to this illness, and, consistent with the three-indicator prediction hypothesis, once probable schizophrenics-to-be are removed from the sample the risk for mental illness for the remainder should be no higher than normal.

By comparison with others in the group, the five children targeted by this study seemed quite impaired. They showed a variety of soft neurological signs. They couldn't write very well because they lacked the fine motor control for manipulating a pencil. They also showed a strange inability to flip their hands alternately from front to back in their laps. Some did it with one hand, but then couldn't do it with the other. This could tie in with recent findings of hemispheric dysfunction in schizophrenia, since there is some evidence that a portion of schizophrenics may have inadequate lateralization of brain function or perhaps lesions in the left hemisphere, leading to confusions in hand dominance (Buchsbaum 1977).

Emotionally, the five looked abnormal at both 4 and 7 years of age. Schizoid behavior, often thought to precede schizophrenia, was found, but with a broadened definition. Usually the terms refer to withdrawn shyness and dreamy behavior. In this case, it also included unstable and irritable emotions. Dr. Hanson explained that shyness alone is normal and no indication of future mental disorder. Thus, he wanted to include the traits that contribute to the "stormy" personalities often seen in prepsychotic children.

One of the five, for instance, was highly distractable, overactive, and silly at age 4. He tended to have an explosive temper and showed poor eye contact with adults. At 7, he was both withdrawn and hostile, high-strung and nervous. He had a variety of abnormal, repetitive movements, and although he was good at math, reading, and spelling, he could not pay attention to the test and was rebellious in that situation. He had many fears.

An indication of the psychic life of these children was the fantasy told by one little girl who was asked to draw a picture and tell a story. She told the tale of a boy who shot his mother, shot himself, prayed his mother back to life, and then buried her alive.

"That's not the kind of thing you hear from most seven-year-olds," said Dr. Hanson. "It sounds like the chaotic thinking you find in adult schizophrenics."

"Pay Attention Please!"

Other studies have searched for hidden deficits in more fine-grained behaviors. Some of the best results so far show that children at risk lack some ability to focus attention when compared to normal groups. The attentional defects are subtle indeed, not at all like the gross deficiencies seen in adult schizophrenics, but they are important enough to impair a child's developing competency, according to University of Minnesota psychologist, Dr. Norman Garmezy, a recognized authority in risk research.

The findings include, for instance, a deficit in visual attention, or perhaps in the ability to process visual information, so that the child makes many errors in detecting the numbers of playing cards flashed on a screen. A deficit has also been found in reaction time, with the at-risk group taking longer to respond to a signal. Finally, there seems to be a problem with sustained attention and vigilance, wherein the children of schizophrenic parents have greater difficulty in picking out a relevant signal from background noise.

The particulars of this research vary from study to study. Some investigators find problems with reaction time, for instance, while others do not. But the evidence all points in one direction, toward a reduced capacity to focus and sustain attention, similar to but less obvious than the deficits found in symptomatic schizophrenics.

Typical of the best research is the work of a group led by Dr. Lois Erlenmeyer-Kimling, a psychiatric geneticist with Columbia University and the New York State Psychiatric Institute. The children, aged 7 to 12, were asked to press a button when they saw two identical playing cards flash successively on the screen. Meanwhile, the children were challenged with distraction in the form of irrelevant numbers read through earphones. Occasionally, two playing cards would come up with the same number but different suits, in which case hitting the button was counted as an error. The children at risk made more errors of this sort than did those in the control group. They also missed identical cards. But their major source of error was even more striking. They hit many of the "irrelevant" trials, in which the two cards were not alike in either number or suit. This suggests to Dr. Erlenmeyer-Kimling that they were responding to irrelevant cues which are screened out by most people, a conclusion consistent with the theory that schizophrenics may have a sensory overload due to inadequate filters (Erlenmeyer-Kimling and Cornblatt 1978).

Further insight came when Dr. Erlenmeyer-Kimling applied signal detection theory to the data to find out whether the defect was due to sensory processes or to some problem with motivation and attitude. She found no differences in motivation between the two groups. The gaps in ability appeared to stem from the one group's inability to discriminate sensory cues (Rutschman, Cornblatt, and Erlenmeyer-Kimling, 1977). It seems that somehow the brains of at-risk children were not processing sensory information quite right.

In other studies of the same children, the New York team found neurological problems with fine motor coordination also. The at-risk children had on the whole more difficulty with right-left distinctions, rapid alternating movements, and ability to wink. In addition, they showed some defects on the Bender-Gestalt and Lincoln-Oseretsky tests, which assess both attention and fine motor control (Erlenmeyer-Kimling 1975). The findings taken together suggest that *at-risk children have developmental lags in neurological maturation as well as a defect in focusing attention.*

Most important of all, it seems as though poor scores for the at-risk children are due to excessively poor performance by a few children, some 10 percent of the risk group. Preliminary evidence suggests that when that 10 percent is removed from the data, the remaining group looks normal. If this is true, that both neurological and attentional lags are coming from the same subset of children, it offers hope again that scientists may achieve a long-sought goal, isolating a special biological subgroup of individuals prone to schizophrenia.

NATURE OR NURTURE—AGAIN

But are these deficits inherited? Many researchers believe so, but they lack evidence to prove the point. Few genetic studies have been done in this area. One of the adoption studies done in Denmark, however, suggests that the attentional deficits could be due to rearing by a schizophrenic parent, with whom the child has no genetic relationship. Referring to a deficit in reaction time, the authors of that work state: "Our data provide but a single conclusion. Individuals who are reared by a schizophrenic parent exhibit by virtue of that experience a decrement in reaction time performance that is not shown by individuals who share only a genetic endowment with the schizophrenic parent and are adopted away" (Van Dyke, Rosenthal, and Rasmussen 1975).

That startling statement probably will not be the last word on the subject for several reasons: (1) The test used in Denmark is not the same as those used in the U.S. studies, so results are difficult to compare. (2) The people tested in Denmark were adults, not children, and they were adopted away at a relatively late age, when they were 7 to 10 years old, which does not adequately separate genetic and environmental effects. (3) A Canadian study of a few adopted-away children of schizophrenic mothers suggests they carried an attentional deficit into their foster homes (Asarnow et al. 1978). Five children in that study were found to have problems on three different kinds of attention tasks, and four of the five were the genetic offspring of schizophrenic mothers. Since they were several years old when adopted, however, the results are difficult to interpret.

An NIMH study in progress in Israel of children raised in the kibbutz, as compared to the usual nuclear family, is expected to shed important new light on the interaction between environment and genes. So far, the schizophrenic offspring have shown a range of soft neurological signs: facial asymmetry, problems with fine motor control, visual perception, left-right distinctions, and auditory-visual integration. These problems showed up in the younger children, but not in those older than 11, suggesting a maturational lag that resolves itself (Marcus 1970).

One of the most intriguing bits of information on attention comes, however, not from the children of schizophrenics, but from the offspring of parents diagnosed as antisocial personalities. It seems, from a Ph.D. study in Minnesota by Lee Marcus, that schizophrenic offspring are not the only children to have trouble with attention. The antisocial group adds its own special brand of trouble (unpublished).

The task in this case was to hold a preparatory set, a readiness to respond, while waiting for a signal. The longer the interval between the words, "Get ready," and the actual signal, the harder it is to hold a set. Normal individuals have no trouble with a 1-second interval; but 15 seconds is difficult. It is even more difficult when the intervals are irregular, so that the subject never knows from one signal to another how long he has to wait. Three groups, plus normals, were tested with the following results: At first the children of schizophrenic, antisocial, and depressed parents all did poorly, compared to normals. Then Marcus began to vary the experiment. First, he told the groups how long they would have to wait on each trial so that the irregular interval could be anticipated.

The children of depressives immediately improved their performance and scored normally. The other two groups stayed low. Next, he raised the stakes by letting the children play for tokens and win money. Now the children of antisocial parents did well. They could not profit from information alone, but they paid attention to the gambling. Only the children of schizophrenics remained impaired under all conditions.

"I think there is a real and fundamental deficit in these children," said Dr. Garmezy. "It suggests that they have difficulty in focusing, which could create a cycle of disadvantage, with failure leading to failure. Poor attending may leave the child greatly vulnerable to stress and predisposed to schizophrenia, since focused attention is necessary for the development of competence, an attribute that is often a hallmark of freedom or recovery from psychopathology."

A Boston study (Garmezy 1978) indicates that the deficit starts very early. Fifty children with psychotic mothers were tested at 1, 3, 5 and 6 years of age. They showed possible attentional deficits at age 1, probable deficits at 3, and clear deficits at 5, with the Continuous Performance Test, a risk-reward test used by many, including Erlenmeyer-Kimling, and the Embedded Figures Test. The mothers of children in this study were diagnosed to have schizophrenic, pyschotic depressive, and manic-depressive illness.

Some Over-react

Along with the attentional deficit, there exists in the children at risk an apparent tendency to over-react to stimuli such as sounds. The over-reaction is measured as a physical effect called the galvanic skin response (GSR) or electrodermal response (EDR). These are sweat reactions in the hands. Whenever the body's autonomic nervous system is responding to some stimulus, whether it be internal feelings or some external sensation, the sweat glands open up and release moisture, picked up by electrodes as increased conductivity. The GSR is a common method of measuring autonomic reactions such as the fight-flight response to danger, and the greater the amplitude in GSR, the greater the autonomic response and the more "charged" the body is in its reaction.

Children of schizophrenics seem to have more amplified responses than do other children, indicating some hyper-responsivity to sensory stimuli. This effect was first picked up 15 years ago by pioneers in the field of risk research, Drs. Sarnoff Mednick and Fini Schulsinger, working in Denmark (Mednick and Schulsinger 1974). The GSR response was the first sign in apparently normally functioning children that the offspring of schizophrenic parents might have some biological markers predisposing them to later disease, and for years the Mednick work stood alone as the only evidence of that. Now, with research at the University of Rochester, the Danish work is being partially confirmed.

In 1962, Mednick and Schulsinger measured the GSR and a variety of other behaviors in 207 offspring at risk. For controls, they had 104 children of normal parents and they pledged to follow the 311 children for 20 years through the age of risk for schizophrenic breakdown. (They picked 15-year-olds for the original test, to increase the chance that Mednick and Schulsinger would still be alive when their subjects completed passage through the age of risk). In the years since then, 20 individuals from the high-risk group have had psychiatric breakdowns, and 8 or so have a definite diagnosis of schizophrenia. Thus, expected breakdowns based on biological findings in normal children can be compared against actual breakdowns in adults.

Autonomic sensitivity was measured with a loud blast of sound in the ears. In addition to that, the Danish group also tested conditioning responses and paired the loud blast with a bell tone to see whether the at-risk group would over-react, not only to the noxious sound, but to the bell also. In several ways, the at-risk children responded more fully. They were faster to react; their response was more amplified; and recovery came more quickly, indicating, said the authors, "a volatile autonomic nervous system that is easily and quickly aroused by mild stress." They also generalized the over-reaction to the bell.

There were other differences between the two groups. The risk group, for instance, had more children rated by psychiatrists as having poor personal adjustment. But it was the galvanic skin response that gripped the attention of Mednick and Schulsinger (1974). They made no claims that the sensitivity was due to genes; it might also have been caused by birth complications and family rearing. The mother's influence "has not been benign," they wrote. "The child sees her as scolding and unreliable . . . this difficult environment has been imposed upon (or perhaps has been responsible for producing) a child whose autonomic nervous system is highly labile, reacting to threat abnormally quickly and with abnormal amplitude. To make things still more difficult, reactions are not specific, but overgeneralized. This serves to broaden the range of stimuli that are adequate to provoke this sensitive autonomic nervous system . . . . In school he seems to react to excitement by withdrawing . . . he is beginning to learn to escape from autonomic arousal by drifting off into idiosyncratic thought."

The Greater Liability of the Preschizophrenic

Who then were the individuals who showed liability for a psychotic break? They were more extreme on virtually every measure of autonomic sensitivity than were the individuals at risk who have remained well. Electrodermal measures, in fact, sharply differentiated the sick from the well groups, although both were offspring of schizophrenic parents. Particularly striking is the fact that the sick individuals showed no signs of being able to habituate to stress. Whereas the well group became accustomed to stress over time and their latency of response lengthened, indicating less physical reaction, the latency of the sick group actually shortened, suggesting increased irritation and no habituation. Nor did the sick group lose that response in extinction trials, but held on to the end, long after the well individuals and control group had lost the conditioned response.

If this autonomic response is so critical, how then does it lead to the behaviors called schizophrenia? Mednick theorized that the preschizophrenic, cursed with an over-reactive physical response, would learn to protect himself by avoiding stimulation through such methods as irrelevant thinking and passive withdrawal. He would be aided in this process, said Mednick, by the unusually rapid recovery of the autonomic response, a reward for avoidance. Mednick's theory has few advocates today, and the rapid recovery data have not been confirmed, but his major findings of autonomic sensitivity appear to be a genuine phenomenon.

An important longitudinal study (Salzman and Klein 1978) of children in Rochester, New York, has now confirmed the central findings of autonomic over-reaction. The recent analysis was done on 46 children, half of whom are at high risk for schizophrenia. Recovery time did not match the Danish findings, but increased amplitude did. Given a loud rush of white noise in the ears, the high-risk children showed a much greater response than did the other group.

"It was a very robust difference. The two groups of children hardly overlapped. It's no mistake; they really are different," said Dr. Leonard Salzman, co-author of the Rochester work.

But the Rochester children also showed something else, a rather paradoxical finding that adds subtle new evidence to the theory of preschizophrenia. The at-risk children did not become as generally aroused in the experimental situation as did their controls. As the testing progressed, the baseline GSR of the control children began to rise, indicating that they were becoming generally mobilized, not in response to a single stimulus, but in reaction to the entire test procedure. The at-risk children, by contrast, showed some elevation in baseline, but not as much.

Dr. Salzman postulated that the rise in baseline values is an adaptive response of the body to a situation that demands attention. The normal reaction, then, is to mobilize the autonomic nervous system in a generalized way. Perhaps in the absence of such mobilization, the reaction to a given stimulus will be greater, simply because the body is not prepared to cope physically; and, the greater reaction may screen out overstimulation. More evidence that such an effect exists in high-risk children came from another part of the study that measured heart rate. Again, the base heart rate of control children climbed higher than that of high-risk children, indicative of lesser autonomic arousal in the second group.

Divergencies Between Data and Theories

But there are problems with the GSR data. Not all the evidence comes out the same way. The New York work by Erlenmeyer-Kimling, for instance, has failed to find any difference in GSR response in high-risk children. This failure to replicate Mednick's findings is causing many scientists to be more cautious than they otherwise would be in interpreting results. Also, the GSR data are nearly always different from one study to the next. One group of children may show increased amplitude; another may show habituation; a third may show latency or conditioning differences. This variety in measurements presents special problems because researchers don't really know what the various aspects of GSR response mean behaviorally, or whether the differences are important.

Nevertheless, the findings do seem to go in one direction; with the exception of the New York work, they all indicate greater reactivity of autonomic function in high-risk children, according to a review of GSR studies by NIMH scientist, Dr. Theodore Zahn.

In tests of adult schizophrenics, Zahn has found that the ANS continues to respond when other individuals have adapted to a stimulus (Zahn 1977). "You don't get exactly the same thing from study to study, but the evidence points one way. There is something there."

He adds, "Normal people will give you four or five GSR responses, then quit. The schizophrenic may stop after five, but come back with another reaction on the eighth or tenth trial. He doesn't habituate as well."

The genetic evidence on GSR points in two opposite directions, leading to complete ambiguity. Studies of normal twins fail to show any heritability at all in most features of the GSR response. Some aspects, such as recovery rate, show genetic control, but the important aspects such as response amplitude do not. Yet, the Danish adoption study by Van Dyke (Van Dyke, Rosenthal, and Rasmussen 1974) produced, by comparison, positive evidence of a genetic component for autonomic hyper-sensitivity. The confusion may be due to fundamental problems with the methodology available for determining genetic factors.

"The schizophrenic may have a genetic predisposition to the overly sensitive ANS," said Zahn. "We can't come to any firm conclusions yet. The genetics of this are really up in the air."

CONVERGENCE OF THEORIES AFTER ALL?

Given that autonomic sensitivity and attentional deficits are real predisposing factors toward schizophrenia, can the two be linked together in a unified theory of disease? The answer is yes, they can. They fit neatly into current biological thinking on schizophrenia and indicate that two streams of evidence may be coming together, leading scientists ever closer to the root processes in this complicated mental disorder.

A major line of speculation on the etiology of schizophrenia postulates that an excess of the neurotransmitter dopamine is overinhibiting certain regions of the brain involved with emotions and attention. The defect could be in many places, including the limbic system, where overinhibition could suppress normal levels of emotional arousal. Such suppression in cerebral response or dampening of the emotional response could be a protective reaction of the schizophrenia-prone individual who is assaulted peripherally by an overload of sensation. But the cerebral dampening does not have to be protective. It could also be a primary effect, wherein the overinhibited limbic system fails to energize and arouse the individual, so that attention does not become focused and the body is not protected against stimuli.

Whatever the case, reduced emotional arousal seems characteristic of the behavior of schizophrenics, and brain-wave studies on adult patients show that they do have reduced cerebral reactions to stimulation. The cerebral response of children at risk is under study and results are eagerly awaited. If it turns out that the offspring of schizophrenics also dampen stimuli even while their bodies over-react, a major new piece of the puzzle will be in place.

REFERENCES

Asarnow, R.F.; Steffy, R.A.; MacCrimmon, D.J.; and Cleghorn, J.M. An attentional assessment of foster children at risk for schizophrenia. In: Wynne, L.C.; Cromwell, R.L., and Matthysse, S., eds. *The Nature of Schizophrenia: New Approaches to Research and Treatment.* New York: Wiley, 1978.

Buchsbaum, M.S. Psychophysiology and schizophrenia. *Schizophrenia Bulletin,* 3(1):7-15, 1977.

Erlenmeyer-Kimling, L., and Cornblatt, B. Attentional measures in a study of children at high risk for schizophrenia. In: Wynne, L.; Cromwell, R.L.; and Matthyse, S., eds. *Nature of Schizophrenia: New Approaches to Research and Treatment.* New York: Wiley, 1978.

Erlenmeyer-Kimling, L. A prospective study of children at risk for schizophrenia: Methodological considerations and some preliminary findings. In: Wirt, R.D.; Winokur, G.; and Roff, M., eds. *Life History Research in Psychopathology.* Minneapolis: University of Minnesota Press, 1975.

Garmezy, N. Children at risk: The search for the antecedents of schizophrenia, Part I: Conceptual models and research methods. *Schizophrenia Bulletin,* No. 8 (Spring):14-91, 1974.

Garmezy, N. Children at risk: The search for the antecedents of schizophrenia, Part II: Ongoing research programs, issues, and intervention. *Schizophrenia Bulletin,* No. 9 (Summer): 55-126, 1974.

Garmezy, N. Current status of a sample of other high-risk research programs. In: Wynne, L.; Cromwell, R.L.; and Matthysse, S., eds. *Nature of Schizophrenia: New Approaches to Research and Treatment.* New York: Wiley, 1978.

Hanson, D.; Gottesman, I.; and Heston, L. Some possible childhood indicators of adult schizophrenia inferred from children of schizophrenics. *British Journal of Psychiatry,* 129:142-155, 1976.

Marcus, J. "Neurological and Physiological Characteristics of Children of Schizophrenic Parents." Paper presented at 7th Congress of the International Association of Child Psychiatry and Allied Professions, Jerusalem, Israel, 1970.

Marcus, L. "Studies of Attention in Children Vulnerable to Psychopathology." Unpublished doctoral dissertation, University of Minnesota, 1972.

Mednick, S., and Schulsinger, F. Some premorbid characteristics related to breakdown in children with schizophrenic mothers. In: Mednick, S.; Schulsinger, F.; Higgins, J.; and Bell, B., eds. *Genetics, Environment, and Psychopathology.* New York: American Elsevier, 1974.

Rutschman, J.; Cornblatt, B.; and Erlenmeyer-Kimling, L. Sustained attention in children at risk for schizophrenia: Report on a continuous performance test. *Archives of General Psychiatry,* 34:571-575, 1977.

Salzman, L., and Klein, R. Habituation and conditioning of electrodermal responses in high risk children. *Schizophrenia Bulletin,* 4(2): 210-222, 1978.

VanDyke, J.L.; Rosenthal, D.; and Rasmussen, P.V. Electrodermal functioning in adopted away offspring of schizophrenics. *Journal of Psychiatric Research,* 10:199-215, 1974.

VanDyke, J.L.; Rosenthal, D.; and Rasmussen, P.V. Schizophrenia: Effects of inheritance and rearing. *Canadian Journal of Behavioral Sciences,* 7:223, 1975.

Zahn, T.P. Autonomic nervous system characteristics possibly related to a genetic predisposition to schizophrenia. *Schizophrenia Bulletin,* 3(1):49-61, 1977.

Section IV.

PERSONALITY

Chapter VIII. Normal Personality

The search for genes in personality resembles the search for an oasis in the desert. On all sides, the shifting sands of changeable traits obliterate the trail, and the far-off image of a genetic source often proves illusory.

Nevertheless, hope of finding such a source has inspired a hardy breed of behavior geneticists to sift through a great deal of sand. Some 25 years of personality testing have gone into the search for a specific component of personality that is reliably identified as inherited.

Some say they have found a genetic source; others disagree. But the cumulative evidence is throwing a new light on personality and particularly on sociability.

SOCIABILITY: INTROVERSION—EXTRAVERSION

By all accepted notions of personality, one would expect sociability to be least influenced by genes. No trait seems more indicative of learning and environment than one's social habits and tendencies. Yet, time and again, a significant genetic component has turned up in studies of introversion and extraversion which, in the parlance of psychological testing, means sociability. Morever, both twin studies and recent adoption studies show the inherited component, so it is a more reliable finding than if one method alone were used.

C.G. Jung's discrimination of "attitudes," as he called them, into extraversion and introversion, coalesced observations about personality around these two useful, and much-used, terms. The British behaviorist, H. J. Eysenck, in his work on the sociability factor, constructed a test that isolated extraversion-introversion as a dimension of personality. That was in the mid-fifties, and his scale has been in the forefront of genetic research into personality traits ever since. It is a highly pervasive trait. Depending on the perspective of the researcher, sociability constitutes one of four to eight fundamental personality traits, and, according to Eysenck, at a gross level of analysis, everything known as "personality" can be broken down into two dimensions—sociability (E and I, standing for extraversion and introversion) and neuroticism (anxiety).

In his original work, Eysenck found a very large genetic factor in extraversion and introversion, as large a factor as is typically found in tests of intelligence, in which, on the average, about half the variation among people on IQ is attributed to genes.

"It would appear that extraversion and introversion have a very considerable hereditary component," wrote behavioral geneticist, Dr. Steven Vandenberg (1967, p. 77). Since then, the genetic evidence on sociability has become even stronger, but the estimated influence of the inherited component has diminished. Later work has not found any aspect of personality to be as strongly inherited as is intelligence.

Extraversion and introversion typically reflect preferences having to do with gregarious as opposed to solitary activities, and all degrees of that dimension.

People who score high on extraversion are likely to answer "yes" to such statements are:

> I like to be with a crowd of people.
> I enjoy going to parties.
> Talking with strangers doesn't bother me at all.
> I'm comfortable around a lot of people.
> It doesn't bother me to get up and speak.

By comparison, those tending toward introversion give positive responses to these sorts of statements:

> I really prefer to do things by myself.
> I avoid large gatherings, if I can.
> I like to work on problems by myself rather than with a group of people.

That this capacity for diversity in social traits should have heritable components fits well into evolutionary theory. The development of sociability and cooperativeness, group behavior in general, appear notably in the record of primate evolution, especially of human evolution. That fact is often overlooked in tales of early man's aggressiveness which fail to mention that man also had to cooperate in groups in order to hunt down large animals and survive in a challenging environment.

Adoption Studies

Dr. Sandra Scarr, at the University of Minnesota, presented new adoption data showing that variations in sociability are genetically influenced. She had this to say about evolution: "It seems to me that relating to other people is a rather basic form of human adaptation. We humans live in social groups and I expect there would have been natural selection against extreme forms of social introversion in that people would not get together and mate. The other extreme of extraversion would have been dysfunctional, too, because we have other things to do than hang around each other and relate."

She added, however, that the genetic evidence, while congruent with evolutionary theory, does not prove that sociability is nourished directly by genes. It could, for instance, be reflecting other sorts of personality traits, such as risk-taking or impulsivity, emotionality, or perhaps just energy level. Possibly much introversion is a form of passivity and much extraversion a form of activity. (Jung's original definition sees extraversion as finding value and reality in the surrounding world and introversion as perceiving more value in the inward than the outward experience.)

TALKING WITH STRANGERS

Several attempts have been made to get beneath the E and I scale, to find out what more specific components it might contain. Usually the efforts land quickly in the shifting sands of personality tests and theory, without much evidence to indicate the genetic source contributing to sociability.

One foray, however, has produced an intriguing hypothesis. Researchers in Texas and Colorado have reason to believe that a genetic factor underlying sociability has to do with the ease and pleasure with which one talks to strangers. This one factor, gregariousness in company with strangers, emerged from a complex analysis of personality tests given to 200 middle-aged World War II veterans who were listed among 16,000 twins in a national veterans twin registry.

Several Twin Studies

Originally, the twins were part of a study aimed at finding out which personality traits make an individual susceptible to heart disease. Dr. Ray Rosenman did the original testing with the California Psychological Inventory, a popular test of so-called normal behavior. Unlike some personality tests, this one carries no implications of mental disturbance. It taps such dimensions as dominance, status seeking, sociability, self-acceptance, responsibility, tolerance, flexibility, femininity and masculinity, achievement through conformance, and achievement through independence.

With this material, Drs. Joseph Horn, of the University of Texas, and Robert Plomin, of the University of Colorado, turned the data around to look at genetic factors. The job they set for themselves was to discover what genetic source might be nourishing several of these personality dimensions. To do that, they had to ignore the scales and go directly to the 480 questions in the test. They compared identical and fraternal twins on each individual question. Items showing the largest and most consistent differences between the two types of twins were deemed to be genetic. There were 50 such items, statements the twins made about themselves which in every case showed the identical twins to be much more alike than the fraternal twins. The 50 items covered a variety of behaviors, but was there a common thread among them? Apparently there was. Analysis of the answers by the Varimex rotation of the principal factor solution produced one major factor which explained 25 percent of the variance. That one factor was best characterized by the question: How do you feel when you meet a stranger you have to interact with?

"Some people are warm and open in that situation. They want to get to know the person and they really like it. Others dread such encounters," said Plomin. "I called this trait gregariousness. It's the quantity type of sociability. It has little to do with qualities such as sensitivity, empathy, and affection—all those things that are important in a one-to-one relationship."

Horn calls the trait "conversational poise" (Horn, Plomin, and Rosenman 1976) because it is limited to talking with strangers. On the basis of its importance in this test, Dr. Horn believes it should be "carefully considered in future studies as the possible genetic foundation for the development of the trait of sociability."

What makes the work of Horn and Plomin interesting is that they had no preconceived notions about which genetic factor would fall out of the test. The 50 items contributed equally to all personality scales of the CPI, and the underlying factor could easily have been emotionality instead of sociability. That would have been much easier to explain physiologically. Emotionality such as quick-temperedness or fearfulness ties directly into the autonomic nervous system; the genetic pathways thus seem more obvious. But how do you explain gregariousness?

"I have no idea what the physiological explanation is and from an evolutionary perspective, I don't think we have to explain it. Selection acts on the total behavior, and to say we have to account for that in physical terms is reductionistic. I'd be willing to bet that the physical system underlying sociability is incredibly complex," said Plomin.

From studies of brain function, however, it is possible to derive a theoretical pathway for sociability. Recognition of strangers constitutes a clear neurological state in the development of an infant, and one could postulate that genes (and the environment acting together) specify a pleasurable or negative reaction to that event, which might then act as a channel of behavior within which later learning and experience operate.

The work of Horn and Plomin has, however, yet to be tested and validated against a new group of twins or adopted children, a necessary step in establishing this conclusion. Even without that, some authorities would accept a genetic base for gregariousness, simply because the extraversion scale is so obviously loaded with statements about talking to strangers. Sandra Scarr, for one, views "conversational poise" not as a component of extraversion but as extraversion itself, as measured by the CPI and the Eysenck scale. In most cases, these tests have turned up evidence of an inherited component.

But how heritable? How strong is the genetic base compared to all the other influences that operate to create social preferences for group or solitary behavior?

It is difficult at best to reach an estimate of genetic strength from twin studies. Twin data are inherently confusing, especially in the arena of personality. Scientists do not know much about the rearing environment of twins and the extent to which parents promote or discourage similarity in their offspring. They may, for instance, treat the twins as equal in some respects and unequal in others, thus reinforcing genetic similarities or causing differences that might not otherwise exist. As a result, inconsistent environmental forces so mix up the twin data that a constant genetic factor is difficult to find.

This fact may explain why some twin studies find dominance to be a highly inherited trait, while others find no genetic input at all. In one study of twin adolescent boys, dominance was a highly inherited characteristic on the CPI (Nichols 1966). Yet, with 55-year-old men, Horn and Plomin found no genetic influence to speak of in the dominance dimension.

These contrary findings could be explained by twin interaction wherein one of the pair becomes a leader while the other takes the role of follower, thus forging separate identities with the same genetic endowment. But the inconsistencies could also be explained by age. Nichols' twins were adolescents, while the veterans were past 50 years. Possibly genetic input to dominance diminishes over time with learning and experience.

DIFFERENT TWIN STUDIES: REARING VS. GENETIC INFLUENCE

Considering such ambiguities in twin studies, many behavioral geneticists have turned eagerly to adoption studies for confirmation of genetic traits in personality. The parallels between identical twins reared apart can be remarkable, as Newman, Freeman, and Holzinger discovered 40 years ago with the landmark study described in chapter II.

In that study, personality presented a different genetic profile from mental ability. First of all, the traits were less similar; secondly, the diversities could not be attributed to any factor, indicating either that important environmental effects on personality have yet to be identified or that they are too random and complex to pin down.

There was one exception, however, to the generally muddy personality picture. One test, the Downey test of will and temperament, showed as consistently strong similarities as did mental ability. This test measures such traits as aggressiveness, speed of decision, resistance to opposition, and impulsivity. Despite their various homes and environments, only 4 pairs of twins were really different on this test; 5 pairs were virtually identical, and the other 10 were very similar.

This was the first study of separated twins and one of only two to use personality test data showing the degree of similarity. The other, a study of 42 pairs, done in 1962 by the late James Shields (1962), reached very much the same conclusion: Identical twins reared apart nevertheless became highly similar, more alike than fraternals reared at home. There was even some tendency in the second study for the separated twins to be more like each other than were their counterparts reared together, indicating the possible diversifying effects of home environments. From the following table it can be clearly seen that the condition of being reared apart made no difference in the degree of similarity among identical twins for tests of either neuroticism or extraversion. Correlations have a possible range from zero (no similarity) to 1.00 (complete identity).

Test	Reared together	Reared apart
NEUROTICISM (Newman et al.)	.56	.58
NEUROTICISM	.38	.53
EXTRAVERSION (Shields)	.42	.61

Correlations for intellectual abilities among separated identical twins are in the range of .75. Four studies have been done on intellectual abilities, all reaching similar conclusions in finding the same kinds of correlations.

Shields also evaluated the twins clinically for signs of psychiatric difficulty and found that almost half of them developed the same kind of trouble. Forty-five percent of the separated twins were alike on a range of childhood neurotic symptoms. Slightly more than half of the identicals raised together (57 percent) had the same symptoms.

Traits that appeared in both twins, if they appeared at all, were enuresis, sleep disturbance, aggressive behavior, and emotional difficulties in adolescence. Symptoms usually not found in both twins were: sleepwalking, tics, overdependency, and crying spells. A range of other symptoms including stomachaches and psychosomatic complaints, anxiety attacks, fears, fainting spells, delinquency, and stammering were as likely as not to be concordant in the two twins.

These findings are consistent with the usual muddy results on neurosis. Some twin studies find no genetic component at all in neurotic symptoms; others find a mild to moderate degree of heritability. These separated twins were equally likely to differ or resemble each other in neurosis and anxiety. Hysteria, however, is usually considered not genetic, and it was discordant in this study. Compulsiveness and obsessional behavior, by contrast, are thought to have a significant genetic component. The Shields' twins were usually concordant for these symptoms.

The diversifying role of the environment is obvious in twins, because this is the only means by which they become different. In twin research the question is: Given the cultural setting of modern America, how much will the environment do to differentiate two people with the same genes?

The rough answer seems to be that it causes a 50 percent difference in personality and psychiatric symptoms, but a much smaller difference in mental ability which is trained by and responsive to the larger cultural framework.

DIFFERENT ADOPTION STUDIES: SOME SURPRISES

But what happens if you turn the question around to ask with adopted children: Are people who have no genes in common made more similar by sharing the same rearing environment?

One of the few adoption studies to have been completed in this country, that of Sandra Scarr, is not only confirming the genetic evidence on sociability but adding important new insight on the strength of the factor (Webber and Scarr 1976).

Critical findings from this study of 240 Minnesota families may startle the parents of adopted children and cause psychologists to take a new look at the impact of rearing on personality.

The families were divided equally into two groups. All the children in the groups were teenaged. The families were much alike in social class and occupational background. The main difference was that in one set of families the children had been adopted, while in the other set they were the biological offspring of the parents. In both sets, the teenagers were compared to their parents and also to their siblings with the following results:

1. There was virtually no similarity between the social tendencies of adopted children and their parents, despite the fact that these children, when they were tested, had been living in the adoptive homes for 15 to 20 years, from birth to adolescence.

2. Biological children did resemble their parents on extraversion-introversion, but not very much. Most of the differences among children were unexplained by any factor. Shared genes and shared personality styles between parents and their natural offspring accounted for only a quarter of the variation. The rest of the variance on extraversion and introversion, 75 percent of it, could not be explained. Children were the way they were for reasons not yet detected.

3. Both adopted and biological children were more like their sisters and brothers than they were like their parents. The

highest similarities of all (accounting for 30 to 35 percent of the variance) lay among biologically related brothers and sisters, who shared not only similar genes, but also a similar age, generation, and uterine and rearing environment. Still, most of the variance was left unexplained.

4. In order of importance, from the least to the most similar, the four groups broke down like this: adopted children and parents (zero correlation), adopted children and their siblings not related by blood (.10), biological children and parents (.25), blood-related siblings (.30 to .35).

"There is something heritable about introversion and extraversion, because clearly the biological families share more than do adopted families," said Scarr. But what that heritable thing is remains unclear. Scarr and Webber attempted without success to break down the global trait into smaller components. Families were tested not only for E and I but also for social distance (how far one keeps others at bay), social potency (how effective one feels in dealing with others), and social impulsiveness (degree of reflectiveness, thoughtfulness, and impulsivity). On all measures, heritability was about the same.

"It was kind of a shame." said Scarr. "We had hoped we might find some differences. Maybe extraversion is heritable because social impulsivity is highly inherited, for instance. But that's not true. I argue that social extraversion-introversion is a basic dimension of personality."

But Scarr cautioned against overinterpreting the importance of genes. "Remember how much of these individual differences we are not explaining. What happens to the other 75 percent? It's not shared in the family. There is a great deal these related family members do not have in common."

More surprising, however, is the almost total lack of any similarity between adopted children and their parents or siblings. That finding brings into question a generation of psychological thinking which supposes that children copy or model their personalities after their parents. In this adoption study no such modeling could be seen, a result that raises in dramatic form new questions about the impact of rearing on children.

This does not mean, of course, that rearing has no effects. A parent can obviously affect a child's personality in multiple ways. A shy parent, for instance, may encourage his child to be talkative while a gregarious parent could overwhelm and silence his offspring. In this way, rearing could have pronounced effects that would not be evident at all in a study measuring similarities.

"Surely there is environmental input to children from parents. But this evidence says there is something there against which you are pushing all the time, and that the final profile this kid has will be the result of what he brings and what you do—both," Scarr added.

The important point is that children apparently do not copy the social tendencies of their parents. Thus, it can be postulated that much of the likeness of biological families is due to genetic similarity.

In light of such evidence, both Scarr and Plomin emphasized the dangers of trying to hammer every child into one sociable mold in the belief that it's good to play with others and bad to be alone. Said Scarr, "When you think about whether all children should be advised to go out and play with other kids all the time, you should take into account that some children get more of a kick out of that than others. A child who's told to play with the gang when he wants to be involved with solitary thoughtful activity may not benefit so much from that advice."

The introversion scale used in these tests carries no implication of abnormality or neuroticism. Those people who measure high on introversion in the CIP and Eysenck tests are just as likely as extraverts to have social relationships. It's large, gregarious gatherings they don't like.

Nor is there any reason to believe that natural selection has acted upon genes in the broad middle range of sociability, although people at extremes might be at a disadvantage.

Scarr believes that, in the very broad middle range, there is probably equal adaptational value for different genes. "I don't know of any evidence that someone who is moderately introverted and who marries an individual who likes relatively quiet people is at any disadvantage compared to someone who likes backslapping and who gets married, only to go out to the bar with the fellows every night."

SOME DIFFERENT TEST RESULTS

But some tests of introversion do carry implications of abnormal behavior, specifically the one in the Minnesota Multiphasic Personality Inventory (MMPI). The MMPI was originally designed to measure psychiatric disturbance, and high scores on this test are viewed as signs of troubled behavior. This kind of introversion is apparently more akin to neuroticism, which in turn is heavily loaded with anxiety. The relationship between this scale and the two so-called "normal" tests of introversion previously discussed is unclear; the MMPI contains items geared to picking up emotional disorder and sees personality from a different perspective.

The point is an important one in studies of heredity. It is even more important in light of some new information that high introversion scores on the MMPI are linked with low reproduction among a group of middle-class, college-educated Minnesota men.

The implication here that personality affects fertility is likely to stimulate some debate. Scientists have wondered for years whether natural selection is operating in contemporary society to give a reproductive advantage to some types of personality, but hopes of getting data on that question seemed slim. Now, there is preliminary evidence that such selection may indeed be going on, according to a long-term study of 900 people who were given MMPIs upon entering the University of Minnesota over 20 years ago.

The group was originally tested between 1951 and 1953 before they married and while they were still young men and women of college age. As a whole, they were very well-functioning, even though many got high scores on one or more of the MMPI scales, indicating some deviancy or abnormality of personality.

In 1976, George Heikens, then a graduate student at Minnesota, followed up the original group to find out how many children they had. Only legitimate children were counted; the existence of illegitimate children was too difficult to establish, making such a count unreliable; therefore, unmarried men and women were automatically counted zero for reproduction. Heikens then determined whether for this group, now middle-aged and married, there was any pattern to their reproduction, on the basis of their earlier personality profiles. There was indeed.

Abnormality on the MMPI was associated with decreased fertility. Morever, the association occurred in a straight linear fashion—the greater the deviation from normal, the lower was the average fertility rate. The men with no abnormal scale scores had an average of 2.64 children each; those with one abnormal score had 2.24 children, while those with two or more abnormal scores had only 2.06 children each. The associations were significant for men, but not for women; that is, overall deviancy did not appear to affect reproduction in women as much as in men, although with a larger sample the figures for women might have become significant. With men and women combined, significant associations remained; thus, it looks as if personality abnormality does have an effect on fertility (Heikens 1976).

When Heikens looked at the specific scales within the test, however, he found even more interesting results. Fertility was not decreased equally on all scales; there was a pronounced pattern to the decline. Introverted men had the lowest fertility of all, lower than men classified as "depressive," "hysteric," "paranoid," "pyschopathic," "neurotic," or "schizoid," the titles of other scales of MMPI. Introverts had even lower fertility than did men who scored high on a test of feminine-interest patterns. The high-feminine men and introverted men were conspicuous in the group for their low fertility, which was due largely to a low marriage rate.

"Psychopathic" men, those who scored high on antisocial items, such as authority conflicts and problems with behavior and law, had on the other hand higher-than-average fertility. They were above average in number of children produced within marriage. Other studies suggest that men with psychopathic profiles also produce more illegitimate children, which would make their fertility differential even greater, said Heikens.

Besides psychopathy, three other classifications with higher-than-average reproduction in this study were: hysteric (sociable, emotional traits); neurotic (anxious, easily fatigued behavior); and schizoid (loose, imaginative, or hallucinatory thinking; erratic behavior). From 90 to 95 percent of the group was married, and for some, especially neurotics, high reproduction could be traced to a very high and stable marriage rate.

Although these four classifications, as a whole, showed higher-than-average fertility, there still was a decline in reproduction among those with more than one abnormal score, where personality disorder may be indicated.

The fact that statistics on illegitimacy could not be estimated, however, makes this study less useful than it otherwise would be in establishing natural selection for personality traits; nevertheless, it is at least suggestive that some male attributes may have a selective advantage over others.

OTHER PERSONALITY TRAITS CONSIDERED

What about personality traits other than sociability? Are they also influenced by genes? What ultimately is the world of possibilities for personality dimensions?

In concrete terms, personality traits are defined by the particular test in use. Each new approach to personality (literally every test) is slightly different from the last one, and there are dozens of tests. As a consequence, there are dozens of different personality studies that cannot be compared.

In the past, much personality work with genetics has been plagued by this lack of consistency. One study finds a genetic component for trait X; another study finds a genetic component for trait Y but not X, and in any case the traits at issue are not quite the same, even though they carry the same names.

"There is no agreement at all on what you're talking about and how you're measuring it," said Dr. Steven Vandenberg (1972).

The pessimism reflected in this statement, with which many behavioral geneticists would agree, is alleviated by recent efforts to consolidate the findings. Clearly, something in personality is inherited. What stands up over the years as a genetic contribution?

According to Vandenberg, three types of traits are good candidates for genetic input—sociability, general activity, or energy level, and a third quality which he thinks of as consistency of behavior, as, for example, in predictable versus impulsive behavior.

Others pick out a similar set of traits but add emotionality (Buss, Plomin, and Willerman 1973). They theorize that four temperamental traits underlie overall personality and provide the genetic source for more complex traits. Their four, named with the acronym, EASI, are emotionality, activity, sociability, and impulsivity. They note, however, that the evidence for a heritable component in impulsivity is not good, so perhaps there are only three temperaments. According to their definitions, emotionality equals intensity of reaction. An individual high in this trait is easily aroused and tends to have an excess of emotions. Whether this extreme is of anger, fear, violent mood swings, or all three is irrelevant. The important thing is arousal. Activity refers, as one would expect, to amount of energy and action, as opposed to passivity. Impulsivity is defined generally as the tendency to respond quickly or inhibit one's momentary reaction and come on more slowly. Proof that this particular constellation of traits is the inherited package has yet to be gathered.

But How Much Is Inherited?

Proof for a set of inherited temperaments would represent a major advance in the field of personality. As things stand now, behavioral geneticists have established a consistent, moderate, generalized input to all of personality, without evidence that trait A is more inherited than trait B.

Many researchers would agree with this assessment by Scarr: "I don't see any evidence of differential heritability for different aspects of personality as we've been able to measure them. You come out with different results according to the age levels and populations you study. Sometimes, some scales are more heritable than others; then the next time, you find other scales that are more heritable. So there are a lot of conflicting results. Nevertheless, people don't usually fail to find some heritable component."

This point is particularly born out now by results from the largest twin study ever done on the genetics of personality (Loehlin and Nichols 1976). This newest and largest study utilizing 850 sets of twins from lists of the National Merit Scholarship Corporation seems to have provided the latest word on traditional twin studies of personality. Essentially, the findings say that genes appear to be contributing equally to all aspects of personality, since there was no tendency whatsoever for monzygotic-dizygotic (MZ-DZ) differences to be greater in one area than in another. Identical twins were simply more alike in all aspects of personality than were their fraternal counterparts. Thus, gaps in similarity between the two types of twins (the usual start of methods for calculating heritability of a trait) remain roughly the same across the board for every dimension measured, from mental ability to energy level, to personality style, to self-concepts and goals, and finally to vocational interests.

"On the face of it, this suggests that all of these categories of traits are about equally influenced by the genes. I submit that this is not a conclusion the average American personality psychologist would have arrived at *a priori* for this set of trait domains," Loehlin commented. His conclusion of equal genetic input to all personality dimensions is one that many behavioral geneticists would accept, not because it is true but because that is all they can see with current research methods and personality tests.

Twin research with personality tests may have had its day. It helped establish genetic contributions to personality, but, beyond that, little more could be learned about critical genes, traits, or temperaments. Horn and Plomin argue that the fault lies in the tests themselves—that scales overlap and often test similar dimensions under different trait names. No wonder the genetic contribution looks equal through all traits. Their aim then is to get beneath the test to individual items and reorganize them in such a way as to highlight some important underlying trait such as "talking to strangers."

The adoption study is another eagerly pursued approach. What does this method say about the genetics of personality? Does it find a generalized genetic input or some pattern of inheritance, heavily loaded in a few areas? The answer has yet to come.

In the Minnesota adoption study, sociability appears mildly inherited, but so do other dimensions, specifically neuroticism and vocational interests. Neuroticism, the only other personality trait so far studied in this sample, carried a significant genetic component of roughly the same importance as sociability. Vocational interests might have been expected to have no genetic input. Yet they clearly did. Biologically related family members were more like each other on artistic, investigative, and social interests, among others, than were family members related through adoption. Again, the distinctions appeared most strongly among siblings, rather than among parent-child pairs.

Vocational Interests a Possible Clue

Siblings related through adoption had virtually no similarities in vocational interest, while biologically related brothers and sisters were significantly more alike on artistic, investigative, and social scales, three of the six dimensions used in this study done by Harold Grotevant, then a graduate student at the University of Minnesota (Grotevant et al. 1976).

Grotevant used the Strong Campbell Interest Inventory that classifies interests into Holland's six categories, representing six different ways of engaging or dealing with the world:

1. Realistic interests—liking to deal with things, or be outdoors; includes a range of down-to-earth preferences, from mechanical to naturalistic interests.
2. Investigative interests—scientific and academic orientation; liking to work in a laboratory, write papers, or pursue investigative leads.

3. Artistic interests—liking to go to symphonies and museums; this does not imply artistic ability, but rather interest in that kind of work and activity.
4. Social interests—wanting to be a teacher, social worker, or minister, among other kinds of socially beneficial work.
5. Enterprising interests—business orientation.
6. Conventional interests—such work as office routines.

The standing of an individual on these six scales might be called the "personality" of his interests. His profile, the relative ranking of each interest, would indeed give a distinct cast to the personality. The person who is primarily interested in investigative things, followed by artistic and then social interests, for instance, would seem different from the one who puts social interests first, followed by artistic and realistic interests (Grotevant, Scarr, and Weinberg 1977).

Preliminary results from the Grotevant study indicate that the interest profile has a moderate genetic component. Adopted children do not resemble their parents—correlations were close to zero—while biologically related parents and children were moderately alike (the highest profile correlation was .29), indicating an effect of genes, but not a large one. Genes also apparently have a small impact on profile elevation—the amount of interest expressed in the six categories (Grotevant, Scarr, and Weinberg 1976). Again, as in other adoption and family studies, similarities were highest among siblings rather than parents and offspring, indicating the cumulative effect of genes, plus age and environment.

Surprisingly enough, this work suggests that people do not select mates who are like themselves in vocational interests, at least not the interests concerned here. Correlation between parents was low on all the measures. Nor did they develop similarities over time, not even after 20 years of marriage. This is consistent with a finding by Vandenberg of low assortative mating for personality (1972). It seems that people do not choose their mates on the basis of similar personality and interests.

On the individual scales themselves, there were some intriguing hints of genetic patterns, but the material is too preliminary to draw definite conclusions. The strongest resemblance for any parent-child pair was on the investigative scale for mothers and daughters, thus suggesting a genetic base. Mothers and adopted daughters had nothing in common on this scale, while the biological pairs were frequently alike (.40). Other genetic patterns appeared for mother-daughter pairs on the realistic scale, father-daughter pairs on the enterprising and conventional scales, and mother-son pairs on the conventional.

There was no correlation between father and son that showed up as significantly higher for biological than for adopted pairs. Social-interest correlations were positive for both, suggesting an effect of rearing. Both biological and adopted mother-son pairs showed a significant relationship on the realistic scale. Adopted sons were almost as likely as natural sons to follow their fathers' social interests, a rare indication that some modeling had taken place in the parent-child relationship.

An exactly opposite effect, perhaps negative modeling, was found on the conventional interest scale for fathers and adopted daughters and for mothers and adopted sons. If the parent was conventional in interest, there was a good chance the adopted child was not, and vice versa, although these negative correlations were not significant. Still, even the best correlations were only moderately high—more evidence that while genes have an influence, the major sources of individual differences in personality still remain largely unexplained.

Said Grotevant: "What comes through strongly is that even though you can say there is a heritable influence, it is a very moderate, rather low impact. We can't say what types of genes are involved. We are at a stalemate until we know more about the actual gene action."

In the final report of the family correlations of the Strong scales, Harold Grotevant, now an Assistant Professor at the University of Texas in Austin, suggested a further hypothesis. He wrote, "The consistent adoptive-biological difference for profile contour provides support for the hypothesis that the *patterning* of interests in individuals is at least as heritable as specific interests." He adds: "Any hypothesis of a genetic influence on behavior automatically raises the question of how such influences occur." On the basis of correlations of the family studies with IQ, among other studies, he suggests the possibility that personality orientations exhibited in activity level, temperament, and extraversion-introversion, for instance, bring about the impact of genetic factors on individual interests. These ingredients of personality may not direct, but certainly can influence, one's vocational and avocational preferences.

Grotevant's report suggests, also, that the rate of physical maturation, which is controlled by hormones and, ultimately, by genes, has an effect on a phenotype such as interests, particularly during adolescence. Social experiences of a father and son, for instance, may be similar because of similarity in the biologically controlled process of physical maturation (Grotevant, Scarr, and Weinberg 1977).

What seems to be critical about these studies is the new perspective they bring to the understanding of personality. First in importance is the awareness that genes have any impact at all on such things as sociability and vocational interests. It may be news to most counselors that some of the difference they see among clients in these purely mental and social areas is not just a matter of rearing and environment.

A New Look at the Environment

Second, and no less important, is the odd new light thrown on the rearing environment. The very low correlations between adopted children and their parents are striking. To Scarr it suggests that the rearing environment is not as powerful a force in creating family resemblances as people have thought. "Psychology has placed much emphasis on the modeling that parents provide and on specific teachings that parents give their children about their own interests and their own areas of expertise. But if you look at adopted families, those won't seem very compelling aspects of the environment."

Paradoxically, the genetic studies may ultimately do more to clarify environmental forces than genetic forces. For example, one of the most nagging questions in twin studies has been an environmental one: To what extent does a very close environment heighten the similarities between identical twins? Does dressing alike and playing together, for instance, make the twins more alike than if parents were to make separations deliberately?

In his study of the 850 Merit Scholarship twins, Loehlin addressed such environmental questions with a compelling result. He could not identify any relationship at all between personality of identical twins and such childhood environments as dressing alike, playing together in childhood or adolescence, having the same teachers, sleeping in the same room, or being consciously treated equally by parents. How twins were treated on these measures had no bearing on whether they resembled each other later in life (1976).

"What kind of environment is that?" Loehlin asked a 1973 conference on behavioral genetics. Clearly, it is acting in mysterious ways. He theorized that the environment serves in part to differentiate the twins through some kind of natural and inherent dichotomous process. If one twin is seen as a leader, the other implicitly becomes the follower. If one is called "messy," the other automatically becomes "neater." In this way, with the environment acting to form opposites, twins with the same genes come to be different from each other.

But for the most part, neither Loehlin nor anyone else has yet been able to pinpoint what major environmental forces are acting to shape personality. In spite of genes and family models, people seem to be the way they are for unknown reasons. Environmental forces appear almost random.

As Scarr expressed it: "Maybe what we are seeing is that people select from their own environment those things that interest them, that are compelling, that they can handle intellectually. Each genotype is different in this regard, but most of us live in a pretty rich and varied environment, so there is a lot of opportunity for us to select what we want."

A philosopher of another era might have called the process "free will."

REFERENCES

Buss, A., and Plomin, R. *A Temperament Theory of Personality Development.* New York: Wiley, 1975.

Buss, A.; Plomin, R.; and Willerman, L. The inheritance of temperaments, *Journal of Personality,* 41:513–524, 1973.

Dworkin, R.; Burke, B.; Maher, B.; and Gottesman, I. "Genetic Influences on the Organization and Development of Personality." Presented at 6th annual meeting, Behavior Genetics Association, Boulder, Colo.

Gottesman, I., and Shields, J. *Schizophrenia and Genetics: A Twin Study Vantage Point.* New York: Academic Press, 1972.

Grotevant, H.; Scarr, S.; and Weinberg, R.A. Patterns of interest similarity in adoptive and biological families. *Journal of Personality and Social Psychology,* 35(9):667–676, 1977.

Grotevant, H.; Scarr, S.; and Weinberg, R.A. "Resemblances of Personality and Interest Styles." Presented at 6th annual meeting of Behavior Genetics Association, Boulder, Colo., 1976.

Heikens, G. "Marriage and Fertility Related to Personality in a College Graduate Population." Presented at 6th annual meeting, Behavior Genetics Association, Boulder, Colo., 1976.

Horn, J.; Plomin, R.; and Rosenman, R. Heritability of personality traits in adult male twins. *Behavior Genetics,* 6(1), 1976.

Loehlin, J., and Nichols, R.C. *Heredity, Environment, and Personality: A Study of 850 Sets of Twins.* Austin: University of Texas Press, 1976.

Loehlin, J.C. "Personality: The Genes and What Environment?" Presented at Conference of Prospects in Behavior Genetics. Austin, Tex., 1973.

Miner, G. The evidence for genetic components in the neuroses, a review. *Archives of General Psychiatry*, 29(1):111–118, 1973.

Nichols, R.C. "The Resemblance of Twins in Personality and Interests." *National Merit Scholarship Corporation Research Report*, 2:1–23, 1966.

Plomin, R., and Willerman, L. A co-twin control study and a twin study of reflection and impulsivity in children. *Journal of Educational Psychology*, 67(4):537–543, 1975.

Scarr, S. Social introversion-extraversion as a heritable response. *Child Development*, 40:823–832, 1969.

Shields, J. *Monozygotic Twins Brought Up Apart and Brought Up Together*. London: Oxford University Press, 1962.

Slater, E. The neurotic constitution. *Journal of Neurology and Psychiatry*, 6:1–16, 1943.

Vandenberg, S. Hereditary factors in normal personality traits. In: Joseph Wortis, ed. *Recent Advances in Biological Psychiatry*, 9:65–105, 1967.

Vandenberg, S. Assortative mating, or who marries whom? *Behavior Genetics*, 2(2/3):127–157, 1972.

Webber, P.L., and Scarr, S. The research utility of broad and component measures of introversion-extraversion. *Dissertation Abstracts International*, Ann Arbor, Mich.: University Microfilms, No. 77-07020.

Chapter IX. Selected Abnormal Traits

Part I. Sex Chromosome Defects

In 1968, a Frenchman named Daniel Hugon claimed in court that he couldn't be held responsible for murder because he had two male sex chromosomes instead of the usual one. It was an unusual defense, the first of its kind, and it was based on the argument that men with the XYY chromosome might be biologically driven to crime, unable to control their behavior because of aberrant genes. That suspicion was based on some evidence, most of it anecdotal. Prisons seemed to accumulate XYY men at an unusual rate, compared to their incidence in the population; secondly, the histories of many of these men looked fairly violent.

BEHAVIOR AND THE EXTRA SEX CHROMOSOMES

Could it be that the double dose of Y, the extra male sexual material, might drive men to crime and violence? It has been reported that XYY men are on the average 6 inches taller than XY men. Are they also more criminal? There is also evidence that tall youngsters tend to be arrested at earlier ages than short ones; further, there is evidence of more broken homes in the history of XYY samples (NIMH 1970). How much difference do these factors make?

The French court may have seen some validity to Hugon's defense because it gave him a reduced sentence. Other courts shied away from the XYY defense, for good reason. Scientific ignorance on the subject was extensive. It was impossible to say what behavior, if any, was affected by the extra Y. On the basis of a 2-day conference summarizing the evidence, the NIMH's Center for Studies of Crime and Delinquency reported in 1970: "It would not be possible to say at the present time that the XYY complement is definitely and invariably associated with behavioral abnormalities." The conferees concluded that aggression had *not* been found to be more common among XYY men than among XY controls; yet the genotype could well increase a man's *vulnerability* to the developing socially deviant behavior.

Since then, new research on the XYY and other anomalies, particularly the XXY (an extra female chromosome), is beginning to straighten out some of the lines of evidence. Although the data are still incomplete, the sex-chromosome work could become richly productive to the study of behavioral genetics. It promises to reveal some basic brain mechanisms that underlie the behavior of individuals with these anomalies. Hopes of finding biological pathways between genes and behavior are particularly good with the sex-chromosome work because the presence of the genes can be detected through the sex-linked chromosome. Most behavioral genetics research begins the other way around, with known behavior and unknown genes. That is a much more difficult route for scientists to take.

The Physical Effects

A surprising aspect of sex-chromosome anomalies is their relatively mild effects. People can live and prosper with three, four, even five sex chromosomes, despite what would appear to be an enormous genetic and biological insult. No other chromosome could be that deranged without severe, often fatal results (McKusick and Claiborne 1973, p. 16). A trisomy (three chromosomes instead of the normal pair) anywhere else in the karyotype, or complement of chromosomes, would produce severe mental retardation and illness, if not death. Mongolism (trisomy 21) offers a good example of the usual damaging consequences of having one too many chromosomes. Yet the XYY wasn't even discovered until 1961, and except for tallness there is nothing obviously different about these men.

Men with the XXY condition, known as Klinefelter's syndrome, show more physical effects: because of an inadequate supply of androgens they have underdeveloped testes and enlarged breasts. The XXYs are also tall; that seems to be a result of the extra sex chromosome, whether male or female. And both types are prone to mild mental retardation, although it is possible to have normal intelligence with either XYY or XXY.

The biological protection from severe damage appears to stem from two characteristics of sex chromosomes: First, the X chromosome can be inactivated. Only one X is fully active in a cell at a time; the others appear to be partially, if not completely, switched off. Thus, extra female chromosomes do not have full effects, which is fortunate because the X is larger by far than the Y chromosome and is responsible for carrying 6 percent of the total genetic material, including many genes that probably have nothing to do with sex. Second, the Y is a tiny chromosome, the smallest in the cell, and it seems not to carry any genes except those determining male sexual development. So its presence in double form may not have many effects. A YO genotype which has only the male chromosome never develops. Apparently the loss of genetic material is too great. An XO genotype, on the other hand, does survive. But women with this condition, called Turner's syndrome, have some physical peculiarities such as short stature, a webbed neck, and some intellectual deficits.

The Behavioral Effects: Study in Denmark

Although the physical effects are mild, these anomalies do have important, if subtle, behavioral effects. A recently published study of Danish men confirms that XYY individuals do have a high-crime rate, higher but not more violent than chromosomally normal men in their same environment (Witkin, et al. 1976).

The Danish work has the important advantage scientifically of starting with a completely unselected population. All Danish men born to Copenhagen residents in the years 1944 to 1946 were in the original sample. From that, the authors chose the tallest men, 4,139 individuals representing the top 15 percent for height, to analyze for sex chromosome aberrations. They found 12 XYYs and 16 XXYs, a rate of 3 and 4 per thousand, respectively. In a general population not selected for height, the incidence is about one per thousand for each of the two anomalies (Walzer 1977).

The crime rate (convictions) among XYY men was about 5 out of 12. Almost half had a record for some sort of criminal conviction which, although tentative, is a high rate indeed compared to the normal rate for criminal convictions among Danish men, which is 9 percent.

The group of XXYs also had an elevated conviction rate of 18.8 percent, but that figure is not high enough to distinguish it from normal, given the small number of individuals involved. So XXYs may or may not have more crime; the evidence doesn't say.

For XYY there seems to be little doubt. The extra Y does create some special risk for developing antisocial behavior. Figuring out precisely what that behavior is and why it occurs will represent a major advance in understanding the biology of personality. The first, but by no means complete, explanation the authors explore is reduced intelligence. The XYYs were as a group mildly retarded, and some of their convictions at least could be traced to poor judgement in not getting away from the scene of the crime, a mistake that smarter men probably wouldn't make. As a result, XYY men might get caught more often, which would have the effect of raising their conviction rate. That could explain some of the increased crime.

Indeed, when the authors compared XYY men with XY men of equally reduced intelligence, they found that mild retardation did explain some of the crime, but not all of it. On the basis of intelligence, education, and social class in the normal sample, the number of criminals to be expected was only 2.06. In fact, there were actually five men with convictions among the XYYs; thus a large proportion of the criminality remained unexplained.

"We cannot at this moment explain what we have found on the basis of impaired intellect. That is only part of the story," said Dr. Witkin.

The authors also looked at the *kinds* of crime committed and could find no more aggressiveness or violence when XYY men were involved. Still, the aggression hypothesis has not been rejected. It may be that in the life histories of the men now being examined the scientist will find evidence of enhanced aggressiveness.

How then can these findings be interpreted? More crime, but not more violence . . . antisocial behavior that is not obviously aggressive.

One possibility is that, in the cognitive tests or personal backgrounds of these men, scientists will find something suggesting a problem with control of impulses. That is the current focus of research with the 12 XYY individuals, who have been subjected to a variety of biological, pyschological, and cognitive tests aimed at finding out what an XYY chromosome complement does to the body. The researchers hope the findings will have a link with behavior.

Witkin said the intention of the research is to take a look at the pyschological defense structure of XYY men. All humans have some means of handling feelings and defending themselves against unwanted impulses. Some people repress feelings; others deny them; still others intellectualize and achieve control that way. It may be that the XYY condition interferes somehow with impulse control and allows feelings that otherwise would be suppressed to express themselves. The team is looking for a cognitive trait that may affect rational control of impulses.

Along these lines, the kinds of crime committed by two of the XYY men were rather bizarre. One man was picked up several times for committing buglaries while the owners were at home. Another called in a false alarm on a stolen automobile and waited for the police to arrive.

Another path of investigation is to examine testosterone levels. The aim here is to find out whether (1) XYY men have unusually high testosterone levels and (2) whether that is associated with crime. The answer coming through the individual case studies was unexpected. Yes, XYY men do have unusually high testosterone levels—not abnormally so, but as a group they cluster in the high normal range. No, these hormonal findings do not correlate with crime. The XYY men without convictions were just as likely as the others to have high testosterone (Raul C. Schiavi, personal communication).

Of course, the full story on testosterone and crime has yet to be told. In some studies of chromosomally normal XY men, a significant link between testosterone and aggression and crime has been found. Ehrenkrantz (1974) tested a group of men in prison and found an association between hormonal levels and aggression. The more violent and so-called "dominant" men had the higher testosterone levels.

Infant and Child Studies: New Findings

On the basis of present findings, it appears that testosterone must be considered a factor in the XYY condition, if not in the criminal behavior. Now the job will be to discover how high hormonal levels can be implicated in the mental-emotional makeup of XYY men.

Some of the best clues to that question are expected to come from another NIMH-supported study, this one a long-term followup of newborn children with sex-chromosome anomalies. In this work by Dr. Stanley Walzer of Boston's Childrens Hospital Medical Center, children are followed and tested from birth through grade school, so that the entire critical developmental period is open to view (1975; 1976; 1977).

The XYY children are too young for conclusions to be drawn yet, but the children with Klinefelter's syndrome happen to be older, and important findings have emerged. Five children have passed the first grade, and their cumulative testings have produced this startling result: XXY children have a serious problem with speech and language development. Specifically they have difficulty expressing themselves, finding the right words for things, a condition known as dysnomia. Rather than saying, "I want some milk," for instance, an XXY child is likely to say, "I want some of the white stuff in the icebox." Several of the children also have trouble articulating words and bringing them together into proper syntax. All have problems with word retrieval and all those who passed the first grade have been referred independently for remedial reading by teachers who had no knowledge of the chromosomal anomaly. The IQ scores of these children are within normal range; their problem seems to be primarily with expressive language. Walzer believes that the reading deficit already visible will appear later in the mild retardation often seen in individuals having Klinefelter's syndrome.

Further behavioral studies of the XXY children, numbering 15, turned up less dramatic but still significant results. As a group the children during infancy were less active, more mild in their responses, more withdrawn and more pliant than control children without the XXY condition. (Controls used were children with balanced chromosomal rearrangements which have no known effects on behavior).

According to Walzer, an individual XXY child could not be picked out as being different from normal, but as a group their characteristics do not show the usual personality range. They tended to be quiet, shy children, often very malleable and easy to manage. They had difficulty with distractibility, which in some cases meant an inability to concentrate. Some felt bothered by noise.

A thumbnail sketch of such a child might go like this: in infancy, easy to manage, very amenable to the parents' caretaking and schedules, a little late with the "terrible twos." Because autonomy and negativism come late, the parents are eager for some rebellion by the time the child is 3. In later years, parents continue to see their child as more shy and less assertive than other children. But except for the language difficulty, no one individual is behaviorally out of the normal range.

As with the Danish work, the Boston study group was picked from a large survey of the general population, so biased selection is not a problem. The individuals do not represent any special social class or race, nor prison or mental population. The Boston children were selected from 13,751 male children born consecutively at the Boston Hospital for Women between April 1970 and November 1974. Chromosomal analysis turned up 11 XXYs and 13 XYYs over the years of survey.

Occurrence of either anomaly was not randomly distributed over time. They clustered in groups of three or four at a time, and then for months nothing would appear. One hiatus lasted for 14 months and was followed for four XYYs in 2 months (Walzer, unpublished paper). Walzer feels that a virus may be involved in the chromosomal nondisjunction that produces an extra sex chromosome. Socioeconomic class had no effect on the distribution.

These anomalies are not passed on from parents to offspring. Individuals with Klinefelter's are usually sterile, and those with XYY, while fertile, do not pass on the extra Y; their progeny carry the normal pair. Thus, nature protects the XY.

No Easy Assumptions

There must be caveats to prevent the easy assumption, such as that fostered by a few journalists in the bizarre Speck case, that all tall men with criminal records possess the 47, XYY chromosome abnormality, or that all XYY types are, *ipso facto,* criminal. Few investigators are entirely content with statistical inferences gathered from penal institutions and from hospitals for the criminally insane, in which epidemiological variables must include such factors as broken or disturbed homes, lower socioeconomic class, malnutrition, and effects even harder to gauge, such as intrauterine complications.

Dr. Seymour Kessler points out that it cannot be assumed that maladaptive consequences must automatically follow a chromosomal disorder. He reasons that some XYY males might be predisposed toward social assertiveness, good social adjustments, and economic success because of their increased stature, since in many cases taller adolescents seem to show greater leadership, popularity, and social success during their formative years (Kessler 1975, p. 71).

It should be remembered, too, that the phenotype, the outward manifestations of the genotype, results from interactions between the genotype and the environment. Dr. Saleem Shah, Chief of NIMH's Center for Studies of Crime and Delinquency, aptly observed that the only hypothesis necessary to demonstrate a genetic contribution of psychopathy, among other conditions, is for some genotypes in certain environments to respond by developing a variety of behavior problems more frequently than other genotypes do. There is the possiblity also that in some other environments persons with better regulated and more socially adapted behaviors may emerge from the same genotype as the former (Shah 1976, p. 64).

Stanley Walzer provided genetic counseling to parents of XYY and XXY children identified in his Boston hospital survey. Despite vitriolic attacks from some members of the press and a number of both the professional and lay public, who called his work unethical, Walzer believed that continuous counseling was a necessary adjunct to the research. For instance, identification of the children at birth allowed him to discover a deficit in language and reading development in XXY children which could then be treated, in the hope that the problem would not compound itself. Unaided, the children could get worse, blame their learning deficits on themselves and perhaps become generally retarded.

Walzer believes, also, that XYY-karyotype youngsters can grow up quite normally; never make any trouble; never go to jail. John Money's views on XO, or women with Turner's syndrome suggest that maximizing their verbal abilities and minimizing their mathematical and nonverbal failures can ease immeasurably the behavioral deficits which result from their particular karyotype (Money 1970).

Part II. Psychopathy

"Persons with this disease cannot speak the truth on any subject."
—Benjamin Rush

"They are dominated by a sort of instinctual rage."
—Philippe Pinel (1801)

"People with this disorder are incapable of significant loyalty to individuals, groups, or social values, and are grossly selfish, callous, irresponsible, impulsive and unable to feel guilty or to learn from experience and punishment. Frustration tolerance is low. [These patterns] bring the individual repeatedly into conflict with society."
—Definition of the antisocial personality, in the Glossary of the American Psychiatric Association (1969)

Psychopathy has been defined in such terms over the years—the closest thing in psychiatry to the proverbial bad seed. The term "psychopathy" in use in Europe and on some psychological tests covers most antisocial behavior, from social alienation to criminality and usually, but not always, includes alcoholism and drug addiction. Considering the range of behavior that falls into this category, one's inclination is to dismiss as unlikely any sort of genetic base for psychopathy. But Danish and American investigators, again with NIMH support, have produced evidence that some genetic structure underlies the antisocial personality.

Some of the studies have focused on criminality rather than psychopathy, so it's important to distinguish between the two.

Behavior geneticists as a rule view criminal behavior as a multifaceted phenomenon with many causes, most of them environmental. Some crimes are committed by people classified as psychopathic, but by far the great majority of convicted offenders have not been diagnosed that way. Scientists involved in these studies do not believe the genes cause criminality; they do think genes have something to do with the formation of psychopathy and alcoholism, both of which are associated with crime.

MORE ADOPTEE STUDIES

The evidence follows:

Schulsinger (1972), a co-investigator in the Danish schizophrenia research in the late sixties, applied the same research strategy to psychopathy. As with schizophrenia, he found that "genetic factors play an important role in the etiology of psychopathy." The strategy was to select a group of adoptees with the disorder in question and then track down their four sets of relatives to find out whether biological or adoptive parents and relatives showed a greater amount of psychopathy. Should the condition appear most frequently in biological relatives with whom the child never lived, scientists would naturally conclude that genes play a role in the genesis of the disorder. That picture appeared in Schulsinger's data. With 57 psychopathic adoptees as the basic study group, he found the most psychopathy by far among biological parents. Among biological fathers, 9.3 percent were diagnosed as psychopathic, compared to 1.9 percent of adoptive fathers and 1.8 percent of controls. None of the mothers was judged to be psychopathic, possibly a reflection of diagnostic practice, since men are much more likely to receive that psychiatric label than women, who are seen as having mental disorder of some other type.

Schulsinger's criteria for psychopathy focused on finding a consistent pattern of impulse-ridden behavior that could be directed either against the self or the internal environment, and it included alcoholism and drug addiction. Only 5 biological fathers were diagnosed psychopathic out of a possible 57, so while the percentage gap between the 2 sets of fathers was large, the actual number involved was quite small. Thus, the genetic base does not appear compelling as a cause for psychopathy. Only one adoptive father was psychopathic, however, which makes the rearing effects seem less important than the genetic.

Drs. Barry Hutchings and Sarnoff Mednick reported on a larger study of adoptees, employing records from the Danish adoptee files compiled by Drs. Kety, Rosenthal, Wender, and Schulsinger for their large schizophrenia study (1975). From a file of 1,145 male adoptees, between the ages of 30 and 44 at the time their criminality was ascertained in 1971, the researchers began their initial screening, matching the adoptees individually with a group of nonadoptees also registered as criminal for sex, age, residence, and occupational status of the fathers. Screening out of such items as simply "being known to the police" for whatever reason, and minor offenses such as traffic violations produced 185 adoptees with records of criminal offenses comparable to felonies in U.S. terminology. The normal risk for registration of a male as a criminal in Denmark is about 9 percent. The rate among the control group was about the same. For the study group it was 16.2 percent.

A check of the biological and adoptive fathers of the criminal group revealed that the rates of criminality of the biological fathers was almost three times that of the rates among the adoptive fathers and the fathers of the nonadopted controls, a finding that suggests an association between the criminality of the sons and their biological fathers. With the adoptees, the association appears on both the biological and the adoptive fathers' sides.

To study the question in greater depth and with even more accuracy, a proband group was selected from among the 185 criminal adoptees to include all whose fathers were identifiable and whose biological and adoptive fathers had been born since 1890. The resulting 143 selections in this final group were then matched to individuals who were not known to the police and whose adoptive fathers' ages and occupations corresponded to those of the probands.

Thirty-three (23 percent) of the adoptive fathers of the criminal probands had criminal records, while only 14 (9.8 percent) of the control adoptive fathers were so identified. What's more, other evidences of criminality, such as total length of sentences and number of cases recorded, compared similarly. Among the index cases, 70, or 49 percent, of the biological fathers were found to have criminal records; among the noncriminals, 40, or 28 percent, had criminal records, but the offenses were not as frequent or as serious. As to the mothers, relatively fewer, biological or adoptive, were registered as criminals, but the data were proportionately like those of the fathers.

There was more mental illness recorded for the index probands than for the controls, and, predictably, psychopathy was predominant in the diagnoses. Among the 37 probands and 32 members of the control group who had at least one biological parent recorded in the Psychiatric Register, the difference was chiefly on the side of the mothers. Thus, there does appear to be a genetic component, although a small one. An environmental component appears also, although less strongly, in the criminality of the adoptive fathers. For the majority, however, criminality remained unexplained after taking into account both genes and rearing environment.

The researchers feel that their study was strongly indicative of the influence of genes on psychopathy, but they caution against applying their Danish findings in other situations. They remark, too, on the problem of separating genetic and environmental influences in the adoptive studies, particularly where such care has been taken in matching the adoptive to the biological parental backgrounds (Hutchings and Mednick 1975).

In reviewing the paper by Hutchings and Mednick, Dr. Lee Robins, of the Washington University School of Medicine, St. Louis, commented on several findings which she found of special interest. She observed, for instance, that the biological fathers of adoptees are much more criminal than their sons but that little difference in rates for fathers and sons showed up among nonadoptees; that biological fathers of *noncriminal* adoptees are more criminal than biological fathers of criminal nonadoptees; and, of course, that biological fathers of adoptees are more criminal than the adoptive fathers. She concludes from this finding that the father's criminality is more important in determining that his child will be adopted than that his child will evidence psychopathic behavior.

In pursuing this thought, she wonders why there is less criminality in the second generation and offers as quite plausible the explanation that the father's high rate has resulted from both genetic and environmental factors but that the environmental may have been removed for the sons by adoption. She suggests, too, that any children of these fathers who had obvious physical and mental defects would in all probability not have been adopted (Robins 1975).

And Twin Studies

Dr. Raymond Crowe, of the Department of Psychiatry at the University of Iowa, comments that the first investigations of the possible influence of heredity on psychopathy were twin studies. According to the eight studies of criminal twins which he reviewed, concordance between MZ twins ranged from 36 to 100 percent, with concordance between DZs ranging up to 54 percent and concordance between same-sex twins higher than between opposite-sex twins. Weighting of heredity *vs* environment is, however, nearly impossible, considering the rearing factors involved. Broken homes, emotional deprivation, parental rejection, and psychopathology of deviance all must be taken into account. In view of this, Dr. Crowe considers both the Schulsinger and the Hutchings-Mednick studies as providing strong evidence of a genetic factor leading to psychopathy and, further, evidence of the importance of the environment as well.

Since he was anxious to conduct an adoption study based on followup and personal data collection, Dr. Crowe was fortunate to locate a group of 52 adoptees born to 41 female offenders incarcerated in the Women's Reformatory and the State Training School in Iowa. By checking lists of consecutive admissions to those two institutions and the State index of adoptions, he was able to select probands born between 1925 and 1956, so that they were in or through the period of risk for developing psychopathy.

Ninety percent of the mothers were felons, convicted for check felonies, prostitution, larceny, adultery, and breaking and entering. The other 10 percent had convictions for lewdness, contributing to the delinquency of a minor, and transmitting a venereal disease, for instance. Little was known about the natural fathers of the probands, although reformatory records inconsistently maintained indicated that nine of the mothers had husbands who had served time in the penitentiary. Later, as he investigated the cases more intensively, Dr. Crowe was to find more evidence of assortative mating.

After a control group was selected from the State index of adoptions and matched as nearly as possible for sex, age, race, and approximate age at time of adoption, the initial followup began. The researcher found that the age of maternal separation of the probands tended to be later than for the controls, a difference of no significance statistically but considered significant psychologically when the results of the survey were analyzed later. Socioeconomic status turned out to be almost identical for the two groups, who were closely matched also for broken homes and parental psychopathology. Impossible to measure, of course, was the extent of any rejection or deprivation which might have occurred during the time the probands were with their biological parents.

Eventually, research criteria were tabulated for each group, 21 male and 15 female probands and 20 male and 15 female controls; there was the usual attrition, refusal to respond and residence out-of-state being the principal reasons. Testing included a 1-hour standard-structure psychiatric interview, complete with medical, psychiatric, and family history, and a social history concerning school days, occupation, legal difficulties, and general social behavior. The MMPI, which was self-administered and returned by somewhat less than half of the individuals, received little weight in the final summing-up.

The case histories drawn from followup information and tests were judged independently and blindly by three experienced clinicians, the final diagnosis being made whenever two of the three judges agreed. In order to arrive at a quantitative estimate of psychopathology, each case was rated also by the Menninger Health Sickness Rating Scale. The validity of the interview diagnoses proved to be supported by court records. Six probands received a definite diagnosis of antisocial personality, and one control was diagnosed as a probable. According to the records, seven of the probands had arrest records as adults, four of them with multiple arrests, and two with multiple convictions. Three were considered felons. On the other hand, none of the controls had been incarcerated for an offense, one of the two who had arrests as adults having been convicted only. Six probands had spent time in penal or correctional institutions, three as juveniles and four as adults.

It was interesting to the researcher and his judges that the distribution of other disorders was almost equally divided between the two groups, even in the case of alcoholism, drug dependence, and personality disorders other than antisocial disorders. The Menninger Health Sickness Rating had separated eight of the probands from the others and from the controls as being much more prone to disorder. The two not diagnosed as having antisocial personality were diagnosed as having inadequate personality and schizoid personality. Otherwise, there was no statistical difference between the groups.

Dr. Crowe has theorized about the factors, biological and environmental, which influenced the development of these adoptees, particularly those diagnosed as having antisocial personality. Of the environmental impact, he discovered that age at adoptive placement and length of time spent in an orphanage or temporary foster home showed a clear association with antisocial personality. Five of the six antisocial probands were not placed for adoption until they were well over 1 year old; the sixth, however, was placed at 4½ months. Contrasted with this is the fact that none of the 10 placed late because of late separation from their natural mothers became antisocial. There is no indication, Dr. Crowe added, that the antisocial probands suffered in infancy any unusually bad treatment that the others did not, since both the foster homes and orphanages were considered quite good.

Nor was there any correlation between antisocial outcome and the socioeconomic status of the adoptive homes, although there was some between that outcome and homes broken by divorce or death, or the deviancy or alcoholism of a parent. The question arose as to whether the adoptive parents may have known about the history of the natural mothers and, therefore, whether this might have had a "self-fulfilling prophecy" effect. Dr. Crowe thinks not, since there was no evidence among the other probands of delinquency, criminality, or any other behavior disorder which might have appeared if that influence had had any effect.

There is some regret that records of the natural fathers were incomplete. Those that were available were undoubtedly an underestimate of the whole picture, since only serious offenders were recorded in the reformatory records and the records were not systematically kept. However, a review of the orphanage records of the antisocial probands did reveal that five of the six alleged natural fathers were reported as offenders. Taken all together, it was known without question that 14 (34 percent) of the natural mothers had husbands with criminal records, indicating the mating of like with like, especially in the case of the six antisocials (Crowe 1974).

The Hutchings-Mednick and Crowe studies were among a group of four on family effects on criminality of which Dr. Lee Robins wrote (1975, p. 122): "The time-worn sentimental phrase 'there are not bad children, only bad parents' is finally being translated into a form in which it can survive scientific scrutiny. We have passed the point of blame and reached the point of examining mechanisms and circumstances under which the sins of the father are visited unto the seventh generation."

REFERENCES

Crowe, R.R. An adoption study of antisocial personality. *Archives of General Psychiatry,* 31(6): 785–491, 1974.

Crowe, R.R. An adoptive study of psychopathy: Preliminary results from arrest records and psychiatric hospital records. In: Fieve, R.R.; Rosenthal, D.; and Brill, H., eds. *Genetic Research in Psychiatry.* Baltimore: The Johns Hopkins University Press, 1975.

Ehrenkrantz, J.; Bliss, E.; and Sheard, M.H. Plasma testosterone: Correlation with aggressive behavior and social dominance in man. *Psychosomatic Medicine,* 36(6):469–475, 1974.

Goodwin, D.W.; Schulsinger, F.; and Hermansen, L., et al. Alcohol problems in adoptees raised apart from alcoholic biological parents. *Archives of General Psychiatry,* 28:238–243, 1973.

Hook, E.B. Behavioral implications of the human XYY genotype. *Science,* 179:139–150, 1973.

Hutchings, B., and Mednick, S.A. Registered criminality in the adoptive and biological parents of registered male adoptees. In: Fieve, R.R.; Rosenthal D.; and Brill, H., eds. *Genetic Research in Psychiatry.* Baltimore: The Johns Hopkins University Press, 1975.

Kessler, S. Extra chromosomes and criminality. In: Fieve, R.R.; Rosenthal, D.; and Brill, H., eds. *Genetic Research in Psychiatry.* Baltimore: The Johns Hopkins University Press, 1975.

McKusick, V., and Claiborne, R. *Medical Genetics.* New York: HP Books, 1973.

Money, J. Behavior-genetics: Principles, methods and examples from XO, XXY and XYY syndromes. *Seminars in Psychiatry,* 2:11–29, 1970.

National Institute of Mental Health. *Report on the XYY Chromosomal Abnormality.* Washington, D.C.: U.S. Government Printing Office, 1970. (PHS pub. no. 2103)

Robins, L.N. Discussion of genetic studies of criminality and psychopathy. In: Fieve, R.R.; Rosenthal, D.; and Brill, H., eds. *Genetic Research in Psychiatry.* Baltimore: The Johns Hopkins University Press, 1975.

Schulsinger, F. Psychopathy, heredity and environment. *International Journal of Mental Health,* 1:190–206, 1972.

Shah, S.A. The 47, XYY chromosomal abnormality: A critical appraisal with respect to antisocial and violent behavior. In: Smith, W.L., and Kling, A., eds. *Issues in Brain/Behavior Control.* New York: Spectrum, 1976.

Walzer, S. Genetics, sex research and behavior. In: Money, J., and Musaph, H., eds. *Handbook of Sexology.* New York: Excerpta Medica, 1977.

Walzer, S., and Gerald, P.S. Social class and frequency of XYY and XXY. *Science,* 190:1228–1229, 1975.

Walzer, S.; Richmond, J.B.; and Gerald, P.S. The implications of sharing genetic information. In: Sperber, M., and Jarvick, L., eds. *Psychiatry and Genetics.* New York: Basic Books, 1976.

Witkin, H.A.; Mednick, S.A.; Schulsinger, F., Bakkestrm, E.; Christiansen, K.O.; Goodenough, D.R. et al. Criminality in XYY and XXY men. *Science,* 193:547–556, 1976.

Section V.

INTELLIGENCE

Chapter X—Intelligence

CULTURAL BASES

Throughout untold centuries, a Micronesian navigator could guide a canoe of men across 100 miles of open ocean without a single instrument, not even a compass. His eyes, ears, and nose were exquisitely tuned to the environment. His memory was filled with the details of celestial rotation. With the two together—memory and sensation—working at almost a semi-conscious level, this navigator "felt" his way unerringly across the Pacific to within a few miles of a small island.

Halfway around the world, a European navigator crosses the Atlantic with equal sureness. But instead of using body and memory to find the way, he uses a series of abstract calculations to plot the path from one point to another. A wide range of instruments, based on generations of Western science, have replaced and extended the function of eyes, ears, nose, and skin. Unlike the earlier South Pacific navigator, the modern sailor cannot stand at the helm of his ship and know intuitively where he is on the face of the earth. But he can describe his position more accurately. He knows in an objective, purely mental way where he is and where he is going, and what is more, he can explain it, which the South Seas navigator could not do.

The two methods accomplished the same goal but with totally different kinds of knowledge. Each would be incompetent in the world of the other. Would they be considered equally intelligent?

Anthropologist T. Gladwin asked that question in 1953 when he tried to give some Western intelligence tests to the residents of the tiny Pacific island of Truk and discovered that they were completely inappropriate for the type of thinking that goes on in Trukese minds. It was true that the Trukese navigator might not score well on an IQ test designed for schoolchildren in California or Iowa, but he could nevertheless turn in a brilliant mental performance on his own terms:

"The navigator sets a course by the rising and setting of stars, having memorized for this purpose the . . . knowledge gleaned from generations of observations on the directions in which stars rise and fall through the seasons. A heading towards a given island . . . is set at a particular season a trifle to the left, or perhaps the right, of a certain star at its setting or rising. Through the night a succession of such stars will rise or fall and each will be noted and the course checked. Between stars, or when the stars are not visible due to daylight or storm, the course is held constant by noting the direction of the wind and waves. A good navigator can tell by observing wave patterns when the wind is shifting its direction and by how much.

"In a dark and starless night the navigator can even tell these things from the sound of the waves as they lap upon the side of the canoe's hull, and the feel of the boat as it travels through the water. All these complex perceptions—visual, auditory, kinesthetic—are integrated into a slight increase or decrease in pressure on the steering paddle, or a grunted instruction to slack off the sail a trifle."

Western navigators do these things with computers and a mechanical guidance system. The Trukese navigator "does it all in his head. This is an astounding intellectual achievement" (Berry and Dasen 1974, p. 31).

Intelligence is culture bound. From the moment a child is born, culture is acting upon his mental growth and development, pulling out some qualities of mind and leaving others untouched, undeveloped or perhaps even inhibited.

Westerners grow up in a cultural climate that places strong value on logical, analytical, and especially verbal abilities. The minds that blossom in this environment are highly adept at categorizing the world into many discrete bits of information and organizing this material into abstract systems of knowledge. This kind of mind conceptualizes, theorizes, and explains, and in so doing it gains a high level of control over the physical environment. Such abilities yield high scores on IQ tests.

But they are apparently not the abilities that guide a Trukese navigator over the open ocean. According to Gladwin, the man does not really form abstract concepts for what he is doing. He reaches his goal by using intuitive processes, body knowledge, and a great memory store of detailed information on sea and sky.

There is now the realization that many non-Western, nonurban people have had no reason to develop the cognitive style so emphasized in the West. It's quite possible that the intellectual and conceptual processes so cherished by Western people, particularly Western scientists, are not an inevitable ability of the mind but, to some extent, a conditioned product. By this reasoning, people who don't need those abilities will not go far in developing them.

An Australian aborigine child, for instance, is not allowed during day or night to explore a world much bigger than the circle of light thrown by a campfire. The Tarahumara Indians, inhabiting remote regions of Mexico's Sierra Madre mountains, will sit in groups for hours in silent communion. By comparison, the environment of most American children contains television, and conversation is an integral part of life in most social classes. It's not difficult to imagine that conceptual and verbal skills are more stimulated by one environment than by the other. Yet the aborigines and the Tarahumara Indians may have cognitive processes that Westerners barely perceive.

"Scientific psychology is the product and reflects the characteristics of the mind of Western Man. It has become clear that some restatement of its constructs may be necessary in light of what has been discovered concerning their cultural relativity," wrote S. Biesheuvel in the foreword to Berry and Dasen's compilation of cross-cultural research.

"Measures of intelligence, as indications of the power of the mind, are strictly comparable only within homogeneous cultures. There is no possibility of comparing the ultimate intellectual capacity of different ethnic or cultural groups."

The Problems of Cognitive Research

One of the most interesting problems in cognitive research is the search for intellectual abilities common to the entire species. Are there cognitive structures typical of the human race? Do humans develop mentally in characteristic ways regardless of the culture in which they are born? Few questions in science are more difficult to answer. The impact of cultural conditioning in shaping expressions of the mind is so thorough that it creates an almost impenetrable mask over innate abilities. People think differently because of their many disparate sources of cultural information, such as their different languages.

In fact, many of the IQ and aptitude tests currently in use for "tracking" in the public school systems or for judging hopeful high schoolers' qualifications for college entrance are presently under fire. Cognitive abilities which have been sharpened for coping with problems of survival in a lower socioeconomic environment are not developed for answering easily the kinds of questions which usually appear on such tests. Nor could most children of the largely white middle class perform well with tests composed in "street language."

From this, one would assume that intelligence is culturally based, a flexible, changeable characteristic of mind that is shaped according to the needs of the environment. Thus, it would seem to have few genetic determinants.

Genetic Bases

But more than 50 years of study on mostly white, English, and American middle-class children have produced consistent evidence, corroborating some of Galton's theories, that genes play a powerful role in determining intellectual differences among individuals within populations.

In a recent extensive summary of the evidence of the Social Science Research Council, three prominent researchers noted that estimates of innate ability in IQ tend to range from .60 to .80 (Loehlin, Lindsey, and Spuhler 1976). In other words, 60 to 80 percent of the variance in the test scores of school children can be attributed to genetic factors. The authors concluded that heredity plays a very prominent role among people of European origin in whom "both the genotype and the environment demonstrably influence IQ." But the former tends "to account for more of the individual variation in IQ than does the latter" (p. 233), under present conditions.

Twins Help Find the Answer

Not only IQ but cognitive development in preschool children is strongly under the influence of a genetic blueprint, according to a major longitudinal study of twins in Kentucky (Wilson 1972 and 1975). Until school age, mental growth in children is an unpredictable, almost inexplicable pattern of lags and spurts, with the child shooting ahead in one area of mental skill and falling behind in another, only to reverse the process

1 year or 2 later. Now, the Louisville, Ky., twin study suggests that this pattern may be guided primarily by genes.

The children, 142 pairs of twins, were followed and tested from birth through age 6. The lags and spurts in intellectual growth, as well as the children's profiles of mental strengths and weaknesses, showed a significant genetic component. So far, however, many of the twins are still relatively young for firm conclusions.

The twin method used in Louisville is prominent in studies of behavioral genetics and it needs special explanation when applied to the controversial area of intelligence. Identical twins develop from a single fertilized egg and therefore have the same genes. With some exceptions, the only behavioral differences between them are due to the environment. Fraternal twins, on the other hand, develop from two eggs and are no more alike genetically than any other siblings, all of whom, on the average, share half their genes. These twins, with both different genes and different environments, should be more dissimilar than are the identical twins and that, indeed, is usually the case. The greater similarity of identical twins yields a component of difference which can be attributed to genes. From this slice of difference, scientists estimate the effects of genes on a behavioral feature. The twin method has many problems and even more critics, but the method does give some idea, although imprecise, of what the genes may be doing.

If evidence for the genetic theory of intelligence were based only on twin studies, it would be open to some questions. Fortunately, the data are backed up by adoption studies, another major avenue into behavioral genetics.

Correlation of Adoptees and Parents on IQ

Tests given in the Southwest to adopted children and their two sets of parents, biological and adoptive, show that IQs of the children correlate more strongly with their biological parents than with the people who adopted them. Reporting on this study at the First International Congress on Twin Studies in 1974, Dr. Joseph Horn, of the University of Texas at Austin, reported that the similarity with biological parents persisted even in children given up for adoption at the time of delivery so that no postnatal influence from the natural parents was possible. Morever, the genetic influence became more visible with age. In one study, children 9 years of age and older were more like their biological parents in IQ level than were children 8 or less.

CONFLICTING EVIDENCE AND PHILOSOPHY

So we are left with an apparent paradox. The intelligence measured on IQ tests is to some unknown degree an artifact of culture. Yet variations in IQ among Western children seem to be powerfully influenced by genes. This seeming contradiction in its multiple forms is at the root of one of the most intriguing, confused, and complex issues in behavioral genetics. Divided by the nature-nurture dichotomy, researchers may peer with special skepticism at data that seem to confirm beliefs held by the opposing camp. Evidence supporting an hereditary component in intelligence is sometimes subjected to such harsh scrutiny that some scientists refuse to work in this chaotic field. Worse, the data may be misinterpreted.

After reviewing 50 years of research the authors of *Race Differences in Intelligence* wrote, "We have been concerned privately by the number of instances in which the political and social preferences of the investigators apparently have grossly biased their interpretation of data. Such distortions appear to be at least as prevalent at environmentalist as at hereditarian extremes . . . consequently, any evidence deriving from a single unreplicated study must be viewed with more than the normal caution stemming from statistical considerations." (Loehlin, Lindsey, and Spuhler 1975.)

The key to much of the confusion lies in understanding two characteristics of heritability studies. Genetics estimates are based on populations, not individuals. They tell how people differ on some aspects of mental ability, and they indicate how much of the difference is due to genes. But they say nothing about how much of the trait itself is controlled by genes. There is no good way to isolate the genetic component of any behavioral trait without knowing something about the biology of the genes in question. Usually the biology is totally unknown, and scientists are left with a statistical measure of variation which says something about the responses of people to their environment but very little about the basic causes of mental abilities.

Because of the nature of population studies, it is entirely possible for genetic factors to be a major cause of differences among people. Yet the trait in question may be largely a product of the environment. Take, for instance, an imaginary mental trait called "prison cognition," a certain blankness of mind which affects people in jail. Convicts may differ on this trait according to their genetic constitutions, but the existence of the trait at all is due to an environmental event, namely imprisonment.

To take another hypothetical example, suppose American boys fight more than do Chinese boys mainly because they watch violence on television at the age of 3. Differences among American boys on aggression may be partly due to genes, but it is the environment, not the genes, that causes the overall high level of aggression in American as compared with Chinese boys.

On the other hand, a high degree of genetic determination need not imply genetic invariance. A trait can be completely genetically determined (such as blood type) yet have considerable genetic variance, including that proportion contributed by mutation. However, there are *general* traits that are genetically determined, which have no genetic variance but are environmentally influenced. Human beings are genetically programmed to walk upright on two legs. Variations in ability to walk are thus due to environmental factors, such as birth defects, accidents, or bad health. Generally speaking, the more important a trait was in the evolution of homo sapiens, the more it has come under genetic control. Strong selective pressures tend to reduce genetic variation, so that most people get the same genes. In this situation, then, differences are due to the environment and heritability estimates approach zero, in spite of the fact that the trait in question is primarily under genetic control.

In other words, environmental variation masks the effect of heredity; on the other hand, genes operate differently under a variety of environmental forces. In population research, many scientists do not understand this see-saw relationship between genes and environment, so they argue vehemently that their piece of the elephant explains most of the observed variation.

An environmentalist in Georgia, for instance, might study children from several social classes, from the poor to the rich, and conclude that mothering style is the single best explanation for IQ differences. Several States away, another scientist studies twins and finds that innate ability explains most. Both may be right, but only for the respective environmental settings.

Heritability statistics based on people in one type of environment cannot be used as the only explanation of variations in another setting. Nor can they alone be used to explain differences between groups of culturally distinct people. As several authorities have pointed out, it is not valid to attribute IQ gaps between black Americans and white Americans to genes.

Sources of environmental variation are too great to compare the two groups genetically, and the development pathways from genotype to phenotype are still too rocky and obscure.

It would be helpful, in behavior-genetic research, to fathom the pathways of development between the genotype, the genetic constitution of an individual or a group, and the phenotype, the manifest qualities that are produced by interaction of the genotype and the environment. A beginning is the knowledge that gene action both regulates and is regulated at every point along the developmental path by the interplay of structural and regulatory genes. The same genotypes can produce various phenotypes depending on significant environmental features which affect the expression of the genotype and also upon the range of reaction of the genotype to a given set of environmental conditions (Scarr-Salapatek 1975).

THE INFLUENCE OF HEREDITY

The science of behavioral genetics would not be so complex and confusing if researchers knew more precisely what they were studying. But in most cases, they do not know which genes are "behavioral" nor how they work, nor do they know exactly how "pieces" of behavior should be defined to correspond most closely to a genetic influence.

What actually is verbal ability? What is spatial ability? What is intelligence? The definitions depend on tests which unfortunately do not all measure the same things. Even tests claiming to measure the same ability, spatial skills, for instance, may not actually do so.

Dr. Steven Vandenberg, a pioneer in twin research on intelligence, once tried to find out what 12 different spatial tests were measuring. He wanted especially to find out if there was any consistency from one test to another. The results were disheartening. In some tests, the ability to see perspective was important; in others it wasn't. Some measured two-dimensional spatial visualization. Some measured three-dimensional ability. Others relied on right-left distinctions. On occasion, trivial details such as poor drawings interfered with accurate measurement.

"It's hard to understand what is going on. The tests are far from perfect," says Vandenberg. Now at the Institute for Behavioral Genetics at the University of Colorado, Vandenberg is thinking of constructing his own tests.

In many ways, however, the central question concerning intelligence is a simple one which has been lost in the heat of controversy over details. The question is this: Do genes make a significant contribution to intellectual traits or is the mind, for all practical purposes, a product of its environment? The answer is that heredity makes a difference, a large difference, in intellectual variation among people from the same cultural background.

Dr. Bernard Davis, of the Harvard Medical School, asked that question at a recent meeting of the American Association for the Advancement of Science and ended his address with a warning:

"The real policy question is whether we can safely assume a value of zero for heritability, or whether we should build on the possibility that genetic differences may be significant. If we fail to face this question honestly and dispassionately, we may find ourselves on the dangerous course of building on illusion . . . .

"Pressure for equality alone, divorced from respect for individuality, will move us toward a totalitarian state rather than toward a more profound democracy. Hence those who would minimize the importance of genetic differences may be imperiling the advance of social justice, however deep their concern."

The evidence is hard to deny. Genes generate individual differences in intelligence. Parents by their rearing practices and teachers by their work are not solely responsible for whether a child is bright or dull. Some of the responsibility lies in the gene.

More Evidence From Adoptees

Nowhere is this fact more evident than in recent adoption studies demonstrating that adopted children follow their own genetic course of intellectual development and do not come to resemble the parents who have reared them, at least not to any important degree.

The facts seem incredible from a psychological perspective—that adopted children continue to resemble their biological progenitors despite separation from them so shortly after delivery and despite years of rearing by other parents. Yet, those are the results on children born and reared within a homogenous, white, middle-class environment. (The qualification is important because the findings do not apply equally to children reared out of their original environment, as a study of transracial adoptions is showing. Radical cultural change may bring quite striking shifts in IQ.)

So the findings reported here apply only to adoptions carried out within a single cultural framework. In that case, the genetic component comes through strongly as a factor in determining individual IQ differences.

For instance, in a sample of some 200 adolescents in Minnesota adopted within the first year of life, the new family environment failed completely to cause any similarity between siblings by adoption. Adopted children reared together in the same family were not alike. Correlations among the IQs were zero, just as they would be among any unrelated strangers (Scarr-Salapatek and Weinberg 1977). Dr. Sandra Scarr, of the University of Minnesota, was surprised to find no commonality among the adopted siblings. "I thought we would find something. After all, two teenagers raised in the same family—but they are not intellectually similar in anything except vocabulary."

Adopted children were a bit more like their adoptive parents then they were like their adoptive siblings, but not by much. The major similarity was with the biological mother. Although IQ scores were not available on the biological mother, the Minnesota adoption records had noted her educational level, which correlated quite strongly with the IQ of the offspring, tested 18 years after birth.

Scarr used rank-order analysis to compare the mother's education and the child's IQ, which means that members of the two groups were ranked according to level of education and level of IQ, and then were compared one with the other. The ranking of the biological mother and child was closer than was that of the adoptive mother and child.

Similar results are being gathered on younger children at the University of Texas, where some 400 adopted white youngsters are under study. In this case, the IQs of biological mothers were available. Mean IQ correlations for the separated adopted child and the biological mother were higher than for adoptive mother and child (Horn, private communication). Horn said the data "support the hypothesis of a significant degree of genetic influence upon individual differences."

Adoption did have some impact on IQ, however. The children as a whole were about five IQ points smarter than they would have been if not adopted. Adoptive homes represent an above-average segment of the population, and adopted children in general have above average intelligence, apparently because of the enriched socioeconomic environment of their homes. General IQ among adopted children runs about 106, compared with

100 for the average population. The Minnesota group had a mean IQ of 106; the Texas group, 108. So the rearing environment of the adoptive home seems to have some impact on elevating IQ generally. But within that context, it doesn't seem to matter much which family a child is reared in. His IQ varies independently of adoptive parental IQ.

"Whether the Jones child has an IQ of 110 and the Smith child an IQ of 120 doesn't seem to be related to the family environment of either the Joneses or the Smiths," said Scarr.

Adoption of black children into white families, however, tells a different story. In the 1960s, Minnesota's social climate encouraged interracial adoption. Not only black children, but Indian, Korean, and Vietnamese children as well, were adopted into liberal, well-to-do white families recruited through both public and private agencies. The children now range in age up to 13 and constitute an excellent opportunity for studying the impact of cultural change on IQ.

The University research team headed by Scarr chose 101 families to evaluate. In sum, they have 321 children over 4 years of age, of whom 176 were adopted. Most of the adopted children are black; only 25 are white, and 21 represent other races.

If the black children had not been adopted, they would have grown up in a cultural environment in Minneapolis where the average IQ of blacks is 90. The question, then, addressed by this study was: Will the children, reared in a different environment, show the same low normal IQs?

The answer is "no." The black adopted children average 106 on IQ tests, the same as white adopted children and 15 points higher than they would have been expected to achieve in their original setting. Fifteen points is the average gap in IQ between races in the United States. It is a gap across which contending camps of academics and politicized followers have engaged in heated controversy over the issue of genes and racial intelligence. Heritability statistics gathered in the white, middle-class culture were used to explain the IQ gap, causing a furious reaction among scientists who believe the intellectual difference is cultural, not genetic. This study well illustrates the influence of the environment on the genotype and the value of a setting conducive to elevating IQ skills.

Moreover, the black children placed in final adoptive families early in life, by the age of 2 months, gained even more, reaching 110, a gain of 20 points over expected levels. The 20-point difference between 110 and 90 is crucial for college education.

It represents the difference between those who can enter college and those who cannot.

"If all black children had environments such as those provided by these adoptive families, we would predict that their IQ scores would be 10 to 20 points higher than under current rearing conditions . . . the scores of these children are environmentally malleable" (Scarr–Salapatek and Weinberg 1977). One statistical reservation remains: The adoptive black children may have been atypically higher in IQ than the 90 average IQ of their peers.

Still, IQ appears to respond to gross changes in cultural environment, not so strongly to fine changes in family environment. That seems to be the conclusion from adoption studies. White middle-class children gained five points in IQ, black children gained 15 to 20 points, directly reflecting the greater cultural change of the latter. But what of individual genetic differences so evident in the white sample of adopted children? Does the genetic blueprint continue to show in transracial adoptions, despite the large environmental impact? It does, but not so strongly.

Black adopted children still resemble their biological parents more than they resemble their adopted parents in rank order, but the magnitude of the genetic effect was smaller under these conditions of adoption. Especially striking was a high similarity between children in the multiracial families. Adoptive siblings in these families, regardless of their race or genetic relationship, were much alike in IQ, a totally opposite finding from the white sample.

Something was happening in the multiracial family to generate much similarity between children not only unrelated to each other but racially different as well. One explanation lies in compensatory education provided by the adoptive mother. Scarr notes that the parents, particularly the mothers, made strong efforts to educate the adopted children with special attention and programs. That could have caused similarities to appear that ordinarily do not occur in adoptive homes.

It may seem from this study that Scarr and Weinberg are advocating adoption of black children by white parents for the sake of IQ. That is not their point. "Black children will and should be adopted by black families. What we do endorse is that *if* higher IQ scores are considered important for educational and occupational success, then there is need for social action that will provide black children with home environments that facilitate the acquisition of intellectual skills tapped by IQ measures."

How Valid Is IQ?

IQ is socially important in some ranges and for some purposes. It has great impact on college education, for example, in the range of 90 to 120. But otherwise there is considerable doubt about the importance of the IQ measure. Once the individual is out of the educational system, IQ is all but eclipsed as a factor in economic and occupational success. For instance, the correlation between income and IQ for white men in 1964 was only .30, meaning that IQ explained less than 10 percent of the differences in income. If Jones made $100 a week and Smith made $200, only $10 of the difference was due to IQ, a very slight tendency for those with higher IQs to make more money (Duncan 1968). Most Americans do not function in their adult lives close to a critical threshold where IQ scores make a substantial impact on their lives (Loehlin, Lindsey, and Spuhler 1975, p. 246).

For these reasons among others, such generalized measures of intelligence have lost scientific interest. The acrimony of the argument over the genetics of IQ has exhausted the tolerance of most social scientists. Furthermore, the concept of IQ itself is increasingly felt to be unproductive. It reflects a narrow range of abilities and says nothing about intellectual processes. Nor does it provide insight into the origins of the cognitive skills that enter into the IQ score.

BRAIN HEMISPHERES: EXCITING NEW RESEARCH AREA

But if IQ is fading as a research topic, the same is hardly true for two components of the IQ, verbal intelligence and spatial ability. One of the most fertile areas of current research is the study of these components and their relationship to hemispheres of the brain.

Material generated by these studies has entered popular language through such terms as "left-brained," referring to an analytical, verbal mind, and "right-brained," a nonverbal, spatial, synthesizing mind. The terms reflect the fact that the two hemispheres of the human brain are specialized for different functions. But the terms have created a vogue and precipitated exalted claims of the respective capacities and functions of the hemisphere, so caution should be exercised in their description.

The left hemisphere normally dominates in many language functions, speech and right-hand dominance, functions which may be reversed in the left-handed individual, although not always. The right hemisphere is often dominant for processing musical stimuli, space relationships, tactile sensations, other nonlinguistic stimuli, fine sensory discriminations, and some left-hand dominance.

Intellectually, the left brain may dominate at sequential, detailed thinking. It may be better at analyzing symbolic relationships of all sorts, both words and numbers. The right hemisphere, by contrast, may be more a gestalt thinker, adept at seeing shapes and matching forms. It may be better at synthesizing rather than analyzing and it may think in shapes rather than in symbols. In its most colloquial definition, the left brain reads and talks, while the right brain finds its way home.

Some investigators in split-brain research believe the two hemispheres actually use different processes which could be mutually antagonistic. For example, if the right thinks in shapes and outlines while the left thinks in logical sequences of symbols, these operations might require different neural organization so that, conceivably, one hemisphere could interfere with the function of the other, if the brain were not adequately lateralized into two specialized and somewhat separate parts.

Split-brain patients—those whose connections between the hemispheres have been totally severed so that the two halves cannot communicate—have provided most of this hemispheric information. One unpublished anecdote of right-left thinking comes from University of Chicago neuropsychologist, Dr. Jerre Levy, who has studied split-brain patients:

The patients are confronted on each side with an array of crosses and squares flashed on a screen. If you ask the subject to match the patterns and point to the one he sees on a sheet in front of him, his right hemisphere will dominate the choice because it is superior in visual matching. Frequently, however, the choice is accurate only in overall shape but not in detail. Three crosses or three squares will be correctly identified, but with any combination of that, the right hemisphere is likely to misplace the order of items. If you force the left hemisphere to respond by asking the patient to describe what he sees, the reporting is much more accurate than the matching was. The right hemisphere pays no attention to detail. It extracts the fundamental form that defines a thing, but if there is no meaning, no gestalt, then the right hemisphere makes mistakes. It is not specialized to order nonsense.

Asymmetric function in the brain was obviously a result of evolutionary processes, and there are two theories to explain its occurrence. One hypothesis proposes that man alone has an asymmetrical brain which developed in order to accommodate language. Abstract linguistic, logical thinking was, by this opinion, imposed on the more primitive non-lateralized, gestalt-thinking brain of the nonhuman primates. When new functions took over the left hemisphere, it gradually lost its spatial, synthesizing skills.

A second hypothesis maintains that asymmetry operates down the phylogenetic line, so that all primate brains, and perhaps those of lower animals as well, are adapted for two types of thinking. Linguistic ability, then, developed from a hemisphere already prepared for the type of processing that underlies the analysis of abstract symbols. By this theory, two types of opposing intellectual functions were important in survival, long before man appeared on the scene (Laughlin and d'Aquili 1974, p. 64).

The issue is not settled, but a conservative course to follow is to assume that, unless and until proven otherwise, human evolution did not invent completely new things but built on already-existing processes. Frequently, functions presumed to exist only in man were found later in other primates but in attenuated form. That is now occurring with language as scientists discover that chimpanzees have some of the intellectual processes that underlie linguistic skills.

Scientists have only begun to tackle the multitude of promising leads flowing from brain lateralization research. But already the field is studded with hot, controversial issues and some striking results.

INTELLIGENCE AND SEX-LINKED CHROMOSOMAL ABNORMALITIES

Two findings of major importance concern the effects on intelligence of certain kinds of sex chromosomal anomalies. Women with Turner's syndrome, whose karyotype (a set of chromosomes) is XO and who have only 45 rather than the usual 46 chromosomes, and men with Klinefelter's syndrome, who have an extra X chromosome, are showing opposite kinds of intellectual deficits.

Turner's syndrome women lack spatial ability to an extreme degree. They are quite good at verbal skills and may excel at reading. But they are so deficient at spatial intelligence that they have been referred to as individuals with two left hemispheres. The reason for the deficit is unknown; evidence suggests some dysfunction of the parietal lobe on the right side (Money 1973). That could be due to the direct impact of the genetic defect on brain cells, or possibly to the lack of release of sex hormones at a critical stage in fetal development and/or puberty. Turner's syndrome women lack all sex steroids but can be brought to puberty by the artificial administration of estrogen. They never become sensitive to testosterone, however, which normal women do secrete and use in small amounts. It is thought that testosterone may have something to do with differentiation of the brain leading to spatial skills. Normal women have poorer spatial skills than do men, on the average, and Turner's syndrome women have a pronounced deficit. Whether or not the two facts have anything in common is a source of much scientific interest.

Klinefelter's syndrome, or XXY boys, on the other hand, by the age of 5 show a pronounced deficiency at expressive language and reading. Yet their spatial skills may soar into the upper ranges of ability. Of six XXY boys studied in Boston, five showed a 15- to 23-point gap between verbal and spatial intelligence. Three had spatial IQ scores of 120 or more. Their performance scores compensated for low verbal scores and placed their total IQ within normal range (Walzer unpublished). Again, there is no good explanation for the deficit, but it could be related to the effect of sex hormones on brain hemispheric lateralization and specialization.

The third finding of importance is included here, not because it falls into hemispheric research as yet, but because it relates to the apparent sex difference in intellectual skills. Reading disability, which affects three to four boys for every one girl in families with affected children, is clustered mainly in the male members. Mothers are slightly affected, but sisters of boys with reading disability show no unusual deficiency. A Colorado team, which did this family distribution study, concludes that reading disability is not sex-linked in the genetic sense, but it may be transmitted by a single gene with reduced penetrance in females or it may be mediated through a polygenic system, with a higher threshold for expression in females than in males (DeFries et al. 1978). The male relatives in this study had no evidence of brain damage or mental disease or deprivation, yet the affected children and their fathers showed a generalized depression of verbal reasoning. Not only were spelling ability and reading comprehension impaired, but there was also evidence of problems with arithmetic and reasoning ability. All of these are thought to be controlled by the left hemisphere. "Reading disability appears not as an isolated handicap affecting only word recognition and spelling, but as a general disorder of language ability," the Colorado team concluded.

SEX DIFFERENCES IN INTELLECTUAL PLUSES AND MINUSES

Evidence of this sort shows a distinct interplay between sex and various types of intellectual deficits. But scientists are proceeding slowly with the data because they don't know what it means, nor do they have evidence of neurological differences between the sexes. Nevertheless, the data on these dysfunctions are congruent with information on normal men and women and strongly suggest that something is going on in neurological development that is different for the two sexes.

The main issue in lateralization research is a sex difference between normal men and women on spatial tasks. Females don't do as well as males. On a variety of spatial tests, boys consistently achieve higher mean scores, so that, as a rule, only 25 percent of the girls reach or exceed the average performance of boys (Harris 1978). A number of independent research studies have produced such data, the most recent being a study of Minnesota teenagers, among whom large sex differences turned up for block design and mental rotation tests. The girls were less able to remember and recreate a design with red and white colored blocks, or to rotate an object in their minds. They were also less able to find their way through mazes, but not to such a degree as with the other tests (Carter 1976). While these spatial results may differ in detail from one study to another, they do consistently indicate that females have more difficulty doing such tasks as orienting themselves in space, making right-left distinctions, and doing geometry.

How and why does such a difference come to be? The answers are speculative and, because the subject is politically and culturally tender at present, a source of considerable heat. Some authorities in the field see this kind of research as sexist, especially when causes for the difference are attributed to genes.

One can ignore genes and postulate, for instance, that women are trained from an early age to think differently from men. Sociocultural traditions offer powerful means for shaping the intellect. Girls talk earlier than do boys; they play differently and they are socialized for more verbal interaction; thus they come to rely more on language and develop a different cognitive style that stresses verbal rather than spatial skills. Boys, on the other hand, are more likely to play with things and to explore greater territory, while remaining less articulate long after girls have learned language. Thus, the intelligence of girls may become more bound up with language, while the male brain spends a longer time developing spatial skills.

The sociocultural explanation appears quite reasonable; it probably does play a role in differentiating intelligence along sexual lines, regardless of what the genes are doing. The genetic explanation holds that males, through the long evolutionary past, traveled farther from home in hunting expeditions and were required to develop extra spatial skills to find their way back. On this view, the sociocultural training extends from a genetic predisposition.

Efforts to find the gene that enhances spatial skills in men are producing mixed results. In particular, scientists have looked for a recessive gene on the X chromosome which, like color blindness, would be expressed primarily in men. Men lack the compensatory action of two X chromosomes, so any recessive trait carried by the X would more probably be expressed in men than in women. With two Xs, the female genetic code can blanket the action of a recessive X-linked gene with normal activity from the other chromosome.

The hypothesized gene in this case is unusually beneficial. Most recessive traits are deleterious; this one supposedly confers enhanced ability on some men. The female level of spatial ability thus would constitute a norm and is not considered a deficit.

Distribution of spatial skills within the population supports a recessive X-linked gene theory. The model predicts that only 25 percent of women will reach the male norm, and that is indeed the case for many spatial tasks. But family studies, the more rigorous test of X-linkage, produce conflicting results. Some scientists have found the anticipated distribution for such a gene, others have not (Harris 1978; Carter 1976). The Minnesota study of 250 teenagers found no pattern at all in the families that would suggest a recessive X-linked gene.

A second possibility is that enhanced spatial skill results from the action of sex steroids, especially testosterone, on the brain, either in the fetal state, during pubescence, or both. Release of sex hormones occurs before birth and is an important event in fetal growth, affecting the development of the nervous system. Whether sexual differentiation resulting from such hormonal action includes spatial ability or not is unknown. But it is plausible. There is some evidence in rats that masculinization of the fetus results in an increased ability to run mazes. Also, bits of evidence in humans suggest that an insufficient supply of sex steroids is accompanied by impaired spatial skills (Harris 1978).

The most controversial explanation (which does not exclude any of the others) concerns brain lateralization. Male brains may be more lateralized than female brains, which would indicate increased specialization in the right hemisphere, if not in both. By this means, the spatial skills of men could be protected against the intrusions of linguistic processes, a protection not accorded to females whose language development comes earlier in life, producing a more generalized, verbally oriented brain.

If this is true, women should show less hemispheric specialization when given various visual and auditory tests aimed at different sides of the brain. Indeed, that is usually the result. Women frequently show equal recognition of objects presented in the right and left fields, while men show differential perception typical of right-brained and left-brained thinking. Such data would seem to settle the issue, but for one problem. The tests are not very reliable. They can't be trusted to reflect the actual degree of brain lateralization. How do you tell in a normal individual which hemisphere is doing the thinking on any perceptual test? Except for split-brain patients, the brain operates as a whole, and there is no measure, including handedness, that is a perfect correlate of dominance by either hemisphere. Nonetheless, sex differences continue to turn up in tests of hemispheric specialization. Some explanation is called for.

More reliable are anatomical findings from brain-injured patients. Sex appears to have an important effect on the pattern of disability, so that men suffer more lasting damage when either the right or the left brain is involved. A drop in verbal ability persists longer in men than in women when the left hemisphere is affected, and the same holds true for spatial ability with right hemisphere damage from strokes or tumors. The conclusion: Male hemispheres are more specialized (Harris, private communication).

Finally, the fourth explanation brings in a complex interplay between handedness, sex, and spatial ability. Very high spatial ability can be found in some (but not all) individuals who lack the usual left-hemisphere/right-hand dominance relationship.

Carter's analysis found that ambidextrous individuals, who show little hand preference, are likely to have high spatial ability paired with poorer verbal ability. Those with strong hand preferences, either to the left or the right, showed the opposite pattern, better verbal than spatial intelligence. Ambidexterity occurred far more commonly among males than among females in the study. Carter does not believe, however, that ambidexterity in the male could account for all the sex differences found between men and women on spatial intelligence.

A different sort of finding, intriguing, although it has been challenged by other researchers, is the observation by Levy of two types of left-handed people with different brain organizations. Left-handed people who write normally with hand straight up and down on the paper are mirror images of right-handed people. That is, the hemispheres are reversed in position, with linguistic functions on the right and space functions on the left. Although reversed, these people still have the usual brain/hand relationship, and they achieve the same results on intelligence and perceptual tests as do right-handers. But left-handed people who write in an inverted position, with hand looped over the paper have ipsilateral dominance or same-sided control. The dominant hemisphere is on the same side as the dominant hand. These people may be either very good or very bad at spatial tasks; they don't fall into the middle (Levy and Reid 1976). An inverted hand posture is much more likely to occur among males, although both sexes show equal amounts of left-handedness. Most left-handed women write in the normal way; most left-handed men write upside down (Harris, private communication).

Perhaps males are more vulnerable to distortions in the usual neural organization. It is well known that the male fetus carries higher risks of physiological damage than does the female. Its death rate before birth is higher and the increased physical vulnerability continues throughout childhood. Men and women are equally distributed in the population, only because more male fetuses are originally conceived. It may be that from that same biological risk comes a higher rate of unusual hemispheric organization, and that whenever the usual left/right, brain-to-hand organization is disturbed, it opens the door to unusual skills mediated through the minor hemisphere.

The evidence cited here reflects the productive confusion and energetic enthusiasm of a new field. Exactly what is going on with hand dominance, sexual specialization, and brain lateralization is still unclear, but when the dust settles it may reveal a variety of hitherto unrecognized neuropsychological types whose intelligence is affected by brain organization and the nature of hemispheric dominance. Male thinkers, female thinkers, verbal thinkers, spatial thinkers, left-handed people with inverted writing postures, left-handed people with normal writing postures, ambidextrous people—the types accumulate, all different, all normal, not because they hit the norm but because they bring their specialized skills together to create a full range of complementary traits.

REFERENCES

Berry, J.W., and Dasen, P.D. *Culture and Cognition: Readings in Cross Cultural Psychology.* London: Methuen and Co., Ltd., 1974.

Carter, S.L. "Structure and Transmission of Individual Differences in Patterns of Cognitive Ability." Ph.D. Thesis, 1976.

DeFries, J.C.; Singer, S.M.; Foch, T.T.; and Lewitter, F.D. Familial nature of reading disability. *British Journal of Psychiatry,* 132:361-367, 1978.

DeFries, J.C.; Vandenberg, S.G.; and McClearn, G.E. Genetics of specific cognitive abilities. *Annual Review of Genetics,* 10:179-207, 1976.

Duncan, O.D. Ability and achievement. *Eugenic Quarterly,* 15:1-11, 1968.

Harris, L.J. Sex differences in spatial ability: Possible environmental, genetic, and neurological factors. In: Kinsbourne, M., ed. *Asymmetrical Function of the Brain.* Cambridge: Cambridge University Press, 1978.

Laughlin, C.D., and d'Aquili, E.G. *Biogenetic Structuralism.* New York: Columbia University Press, 1974.

Levy, J., and Reid, M. Variations in writing posture and cerebral organization. *Science,* 194:337, 1976.

Loehlin, J.; Lindsey, G.; and Spuhler, J. *Race Differences in Intelligence.* San Francisco: W.H. Freeman, 1975.

Money, J. Behavior genetics: Principles, methods and examples from XO, XXY and XYY syndromes. *Seminars in Psychiatry,* 2:11-29, 1970.

Money, J. Turner's syndrome and parietal lobe functions. *Cortex,* 9:387-393, 1973.

Money, J., and Alexander, D. Turner's syndrome: Further demonstration of the presence of specific cognitional deficiencies. *Journal of Medical Genetics,* 3:47-48, 1966.

Munsinger, H. Children's resemblance to their biological and adopting parents in two ethnic groups. *Behavior Genetics,* 5:239–254, 1975.

Scarr-Salapatek, S. Genetics and the development of intelligence. In: Horowitz, F.D., ed. *Review of Child Development Research.* Chicago: University of Chicago Press, 1975.

Scarr-Salapatek, S., and Weinberg, R. IQ test performance of black children adopted by white families. *American Psychologist,* 3(10):726–739, 1976.

Scarr-Salapatek, S., and Weinberg, R. Intellectual similarities within families of both adopted and biological children. *Intelligence,* 1(2):170–192, 1977.

Vandenberg, S. What do we know today about the inheritance of intelligence and how do we know it? In: Cancro, R., ed. *Intelligence: Genetic and Environmental Influences.* New York: Grune and Stratton, 1971.

Vandenberg, S. The nature and nurture of intelligence. In: Glass, D., ed. *Genetics.* New York: Rockefeller University Press and Russell Sage Foundation, 1968.

Waber, D.P. Sex differences in cognition: A function of maturation rate? *Science,* 192:572–573, 1976.

Wilson, R. Mental and motor development in infant twins. *Developmental Psychology,* 7(3), 1972.

Wilson, R. Twins, early mental development. *Science,* 175:914–917, 1975.

Wilson, R. Twins, patterns of cognitive development as measured on the Wechsler preschool and primary scale of intelligence. *Developmental Psychology,* 11:126–134, 1975.

GLOSSARY

acetylcholine—A major neurotransmitter released at the nerve ending to excite neighboring neurons. Thought to play in the brain an opposing role to norepinephrine, serotonin, and dopamine which act to inhibit neural excitation.

ACTH—Adrenocorticotropic hormone. A pituitary hormone that stimulates the cortex of the adrenal gland. This hormone is of major current interest as a possible chemical source of behavioral arousal. It is implicated in studies of depression.

allele—A gene variant. Alleles always occupy the same locus on a chromosome. They produce different enzymatic products which perform similar gene functions.

amine—One of a group of organic compounds containing nitrogen. In the brain, amines function as neurotransmitters and are thought to play a central role in mental disorder, as well as in normal behavior.

blastoderm—A disk of cells which forms on the surface of yolk and which becomes an embryo in avian and reptilian species. In mammalian species, the blastoderm lies between the yolk sac and amniotic cavity.

catecholamine—A type of amine, including norepinephrine and dopamine, which inhibits neural transmission. A deficiency of catecholamines in certain regions of the brain is thought to cause depression. The theory, under some current challenge, is referred to as the "catecholamine hypothesis."

cation—A positively charged ion or electrolyte in nerve cells. The four principle cations—sodium, calcium, potassium, and magnesium—are responsible for nervous activity. Their distribution across the membrane changes as the nerve cell alternately builds electrical potential and discharges.

cyclic AMP—A neurochemical that responds to the action of norepinephrine, serotonin, and dopamine. It functions in the nerve cell to inhibit excitation directly whenever these transmitters cross the synapse. Cyclic AMP is of major scientific interest, but its role in behavior is unknown. High levels of CAMP have been associated with a form of aggression in a laboratory mouse.

dopamine—A chemical transmitter of the central nervous system and one of the most abundant. Dopamine inhibits transmission between neurons in many regions of the brain. It is implicated in the genesis of schizophrenia.

DBH—Dopamine-beta-hydroxylase. DBH is an enzyme that converts dopamine to norepinephrine. Scientists are seeking to establish whether this enzyme functions in schizophrenia, with a deficiency leading perhaps to a buildup of dopamine in the brain.

electrodermal response (EDR)—The reaction of the sweat glands in the hands to emotional or sensory stimulation. Increased sweat reactions raise the electrical conductivity of the hands, and that is used as a measure of autonomic arousal.

electrolyte—A solution that conducts an electrical current. In nerve cells, electrolytes provide the basic electrical activity, without which no other nerve function is possible. The major electrolytes implicated in behavior are sodium, calcium, magnesium, and potassium.

galvanic skin response (GSR)—Refer to electrodermal response.

glial cell—The nonnervous tissue of the brain. These cells support the neurons, providing nutrition and carrying away dead and injured nerve tissue. Glia, which means "glue," are frequently the source of brain tumors.

karyotype—A stained chromosome, prepared so that it can be identified. The process of karyotyping chromosomes, developed in the past 20 years, has made it possible to identify and study specific chromosomes.

Klinefelter's syndrome—An abnormality of the sex chromosome which provides the individual with three, rather than two, chromosomes. The individual becomes male, but because of an extra female chromosome (XXY), he has inadequate masculine sexual development.

L-dopa—A chemical precursor of dopamine. L-dopa is best known as a drug for treating Parkinson's disease. It replenishes dopamine in motor areas that are deficient—the probable cause of Parkinsonism. L-dopa may also cause aberrant behavior since the flow of dopamine is associated with flagrant, manic-type behaviors.

Lesch-Nyhan syndrome—A genetic disease, caused by an enzyme deficiency, which has among its symptoms self-mutilation and neuromuscular disorder.

Mendelian trait—A genetic characteristic controlled by a single gene with dominant and recessive alleles. In breeding experiments, the Mendelian trait is distributed in a classic fashion—three offspring have the dominant characteristic for every individual with a recessive trait.

monoamine oxidase (MAO)—An enzyme that functions in degrading pathways to break down neurotransmitters such as norepinephrine and serotonin. Low levels of this enzyme are implicated in schizophrenia. Inhibition of the enzyme produces an antidepressant effect. MAO inhibitors were the first drugs to be used in the treatment of depression.

mosaicism—The condition of having mixed normal and abnormal chromosomes. An individual with mosaicism has various amounts of normal cells and trisomies (cells with three chromosomes) resulting in various degrees of illness.

myelin sheath—The fatty, semifluid covering of nerve fibers. It serves to insulate the fiber and speed the rate of impulses. Composed of cholesterol and other fatty acids.

norepinephrine—A neurochemical that functions in both the peripheral and central nervous systems. In the periphery, norepinephrine stimulates the release of adrenalin; in the central nervous system, it inhibits nerve transmission in various brain regions, including the "emotional brain" or limbic system. It is implicated in depression.

peptide—A level of biochemical synthesis intermediate between amino acids and proteins. A protein is composed of one or more peptides. Some of these protein particles have been shown to have behavioral effects.

pheromone—A volatile chemical used in scent communciation, often thought to be a sex signal, but also used for other purposes, including trail marking and social-status indications.

serotonin—A neurochemical that functions in both the central and periphral nervous systems. In the periphery, serotonin functions as a vasoconstrictor; in the central, it is said to inhibit certain brain regions during sleep. Serotonin is implicated in the sleep process, mood disorder, and psychosis.

Turner's syndrome—An abnormality of the sex chromosome which provides the individual with one instead of two chromosomes. The individual with Turner's is always a woman (XO) since a human individual cannot survive with the male chromosome alone. A Turner's woman has no sex steroids; she is brought to puberty with the artificial administration of estrogen.

Afterword

Part I: Genetic Counseling

Charlotte Dickinson Moore
Division of Scientific and Public Information
National Institute of Mental Health

Wide-ranging as it is, this monograph has barely touched on some disorders which many believe to be influenced, at least partially, by genetic forces. Mental retardation is one of these. Psychosomatic problems make up another. Alcoholism has received little mention, although many people, lay and professional, view it as genetically transmitted in some measure. As we have seen, some scientists see alcoholism and depression as closely related syndromes within constellations of disturbed families. In generations succeeding ours, perhaps scientists will discover genetic factors in some drug addictions.

Certain knowledge about the role of the genes in shaping behavior is still superficial, still fragmented. Scientists can be justifiably satisfied with all they have learned through their elegant research designs even while, with humility, they continue to search. The folk wisdom that psychotic behavior, mental ability, or even sociability "runs in the family" is corroborated by current research. The preceding chapters recount the hard work and excitement of the hunt. But learning how, and how far, and why, genetic input influences behavior and how it interacts with the environment is an ongoing quest.

So far, the genes-to-traits and traits-to-genes pathways are not clear. To a gregarious individual, sociability may not seem as complex as the cricket's mating song. To scientists, the genetic pathway for that personality trait is more complex, but they believe it is there somewhere. The genetic basis of Huntington's chorea is plain. Otherwise, specific gene-linkage has not been thoroughly identified for behavior traits, unless metabolic errors which affect both body and brain are included.

However, the significant impact of the genes in personal and mental development has been indicated in many ways: in studies of twins, mono- or dizygotic, reared together or apart, in similar or widely dissimilar environments; in studies of first- and second-degree relatives of schizophrenics, depressives, and sociopaths; in longitudinal, prospective studies of children at risk; and in painstaking biochemical investigations. Environment plus the genetic component may both contribute in varying degrees to schizophrenia, unipolar depression, and alcoholism.

At the same time, the impact of environment appears less significant than many in the recent past had thought for a variety of political or philosophical reasons. The emphasis of over half a century on environmental variables is now being qualified. Scarr's study of teenagers adopted at birth indicates the strength of the genetic base with its typical correlation of about .50 for mental ability and .35 for sociability in blood-related siblings raised apart, compared with the nearly negative correlation in these same traits between nonrelated teenagers reared from birth in the same home. Interestingly, some twin studies suggest that the same environment may impinge differently on individuals who share the same genetic load. How much of this phenomenon is due to the usual differences in family interactions and how much to internal mechanisms and substances such as, say catecholamines, which alter individual responses to environmental processes?

A few decades ago, the integration of biochemistry with new fields such as behavioral neurochemistry and physiological psychology might have seemed unlikely to all but the most visionary. Today, their interlocking relationship is taken for granted. Each needs the other, the better to understand the interworking of genetic action with other aspects of living.

GENETIC COUNSELING: WHAT, WHY, WHEN?

One of the useful applications of what we are learning about behavioral genetics is to genetic counseling. Exactly what is genetic counseling? It is ". . . a communication process which deals with the human problems associated with the occurrence, or with the risk of occurrence, of a genetic disorder in a family." So begins the definition drafted by an ad hoc committee of the American Society of Human Genetics in 1973 (Bergler 1979). The definition continues, "The process involves an attempt, by one or more appropriately trained persons, to help the individual or family:

1. to comprehend the medical facts, including the diagnosis, probable course of the disorder, and the available management;

2. to appreciate the way heredity contributes to the disorder, and the risk of recurrence in specified relatives;

3. to understand the alternatives for dealing with the risk of recurrence;

4. to choose the course of action which seems appropriate to them in view of their risk, their family goals, and their ethical and religious standards, and to act in accordance with that decision;

5. to make the best possible adjustment to the disorder in an affected family member and/or to the risk of recurrence of that disorder."

Such a definition is valuable in establishing certain parameters of the contribution of heredity as far as that is known for a specific condition, for understanding the risks and the alternatives of producing offspring in a family with a known genetic-linked disorder, and for making decisions formed out of sound information. Further, it disassociates genetic counseling, as a scientific, yet sympathetic procedure, from the stigma still linked with "eugenics."

Contributions from many scientific disciplines are needed to provide answers for persons whose family histories suggest the wisdom of determining the risk to themselves and their offspring, present or potential, of genetic-linked disorders. Prenatal diagnosis for chromosomal and metabolic disorders that strongly affect normal neurological and mental development in the central nervous system has been more successful so far than other diagnoses for which, usually, only a statistical statement of risk prediction can be provided. In "Intrauterine Diagnosis and Genetic Counseling in Psychiatry," one thoughtful commentator, Gilbert Omenn, explains (1976, p. 144): "The 'phenotype' of mental retardation has been sorted on clinical and laboratory grounds into numerous specific etiologic mechanisms, for which specific diagnostic tests can be applied. I strongly believe that similar processes of sorting out of multiple mechanisms will be essential to our understanding and to our diagnostic approach for the other common behavioral phenotypes, including schizophrenia, affective disorder, and perhaps even addictability to drugs."

Because the etiology of psychiatric disorders is not fully known and the risks cannot be fully estimated, these disorders are not conditions for which counseling can be provided with much assurance, according to some experts. On the other hand, there are physical disorders for which genetic counseling is furnished, such as diabetes mellitus, a disease comparable to a serious and comparatively well-defined mental disorder such as schizophrenia, since each is affected by both genetic and environmental factors. Further, both conditions may have several genetically distinct causes, neither is likely to be apparent at birth, and early and late onset forms of each have been distinguished.

Dr. Ming Tsuang draws on the extensive research he and his colleagues have done on schizophrenia, the affective disorders, sociopathy, and alcoholism to add (1978, p. 1466): ". . . doctors can serve patients and their families beyond offering risk estimates: they can dispel mistaken notions about psychiatric disorders, calm needless anxiety, and help those at risk to draw up a rational plan of action based on the best available information."

According to Dr. Herbert A. Lubs, in the Department of Pediatrics at Denver's University of Colorado Medical Center, the delivery of genetic counseling has two primary goals, prospective and retrospective. "Prospective counseling" is that delivered to someone in the reproductive age group *before* the birth of an affected child, possibly even during premarital counseling. Prospective spouses of persons whose families have a history of mental illness or chromosomal anomaly will want counseling concerning the prospects of the future partners and potential progeny; possibly, too, some of the young adults seeking counseling will be among those who should be tested as potentially heterozygous carriers for diseases such as Tay-Sachs or sickle-cell anemia.

Tsuang's suggestion that doctors can serve in ways other than by offering risk estimates suggests "retrospective" counseling provided *after* the birth of an affected child or to patients themselves. There are "mistaken notions" which should be dispelled and which can be dispensed with fairly easily with sound counseling. The variability of genetic factors and their interaction with environmental forces mean that harmful genes rarely have a 100 percent genetic influence, particularly in psychiatric disorders. And often there are carriers of these genes who suffer no bad effects themselves. Some patients may request genetic counseling if they suffer sporadically and retain enough self-awareness to want to realize what the illness can mean to themselves, their families, and their offspring. Families may seek help from a psychiatric genetic counselor if many family members have had the same disorder or if a few have exhibited severe or chronic illness. Occasionally, particularly in the case of an affective disorder, families and the patient-members can be counseled together.

The "needless anxiety" Tsuang cites can be allayed for many when the risk estimates are presented. For others, alarm can be lessened by the knowledge, for instance, that for many whose diagnosis is manic-depressive disease, lithium carbonate has been of great benefit. Some schizophrenia and schizophrenia-spectrum patients have had their disorders eased considerably with drug therapy. And in the case of sex-chromosome abnormalities and their relation to sociopathies, most men with XYY karyotypes are *not* psychopathic criminals, and most persons with XYY, XXY, or XO karyotypes do *not* suffer mental impairment. Further, the less socially threatening, special learning problems can often be compensated for by special training.

Many of today's young parents wisely reply "just so it's healthy" to the question, "Well, what do you want, a boy or a girl?" The fear of producing a retarded or deformed child, or one who will develop mental illness, whether the cause be congenital, drug-related, or accidental from birth trauma, occurs to most happy couples at some time during pregnancy. According to Gilbert Omenn, approximately 3 percent of all live births will be severely retarded or will have some serious congenital malformations, and the risks may be much higher in a family having a child or other relative already affected (1976). Couples whose fears have some basis in such a family history should receive as precise a diagnosis as possible from a genetic counselor.

Risk prediction is imprecise at best. Even detailed family history, radiologic studies, chromosome karyotyping, enzyme assays, and the relatively recent amniocentesis cannot give complete assurance that any given infant will be free of all defects or birth anomalies. Further, amniocentesis is still considered by some to be an investigative procedure whose possible hazards are not yet fully known. But with all these provisos and seeming dissuasions, genetic counseling, within the last two decades especially, has helped numbers of people to understand the genetic and medical aspects of many disorders and to make their own life choices in the light of this understanding.

A look at the imperfect present and the hoped-for future in genetic counseling for the psychotic conditions—schizophrenia and the affective disorders—and for sociopathy and certain sex-linked chromosomal disorders, may clarify the possibilities of this growing discipline.

Schizophrenia

Lubs has pointed out (p. 9) that schizophrenia, estimated to occur in 1 percent of the population, is unusual in that counseling is more commonly requested because of the concern of siblings and children of schizophrenic relatives that they will themselves develop the disease than because of the risk for subsequent children. Little prospective counseling can be done, he notes, until a biochemical abnormality is identified.

Evidence for the presence of genetic factors, interacting with the environmental factors, comes, according to the work discussed earlier in this volume, from:

1. Family studies, which have shown significantly higher rates of risk among first-degree relatives of schizophrenic probands than among second-degree relatives and somewhat higher rates of risk among those than among persons of similar age, sex, and socioeconomic levels in the general population.

2. Twin studies, which produced far greater rates of concordance for schizophrenia in monozygotic than in dizygotic twins and showed more frequent occurrence of schizophrenia in full siblings of affected probands reared either together, with, or apart from them than in half-siblings who shared the same environment with them.

3. Adoptive studies, which showed a higher incidence of schizophrenia or "schizophrenia spectrum" disease in the biologic offspring of affected individuals, regardless of the adoptive environments of those offspring.

The genetic mechanisms of schizophrenia are still an enigma, despite the many research approaches and the theories which have evolved from each approach. Is the disposition monogenic, polygenic, heterogenic? Possibly further catecholamine research will supplement the findings already gleaned from the twin and adoption studies, and perhaps a synthesis of these approaches may provide more solid conclusions.

In the meantime, genetic counselors must determine the most effective means of reaching the population most in need of counseling. Erlenmeyer-Kimling shows the magnitude of that problem: Of those who will eventually be diagnosed as schizophrenic, 80 percent have neither an affected parent nor an affected sibling; most parents whose children will eventually become schizophrenic cannot be recognized beforehand nor can they receive counseling about births subsequent to the affected child, because usually their families will have been completed long before any child shows a sign of the illness (1976, p. 126). Some siblings of schizophrenics will have finished their childbearing before the affected member manifests sufficient symptoms of the condition to be hospitalized.

A few siblings will seek counseling before deciding on parenthood; the children of schizophrenic parents may well want advice also. The news for these people will be easier to accept than for relatives of Huntington's patients, whose risk is 50 percent if the disease occurs within their nuclear family. The preschizophrenic cannot usually be identified as being at risk until after a breakdown, regardless of relationship to a known schizophrenic, hence most genetic counseling for persons with this disorder is after the fact.

Genetic counseling is highly desirable, nevertheless, in the case of schizophrenia. Despite the findings that, unless a child already has a "genetic load" for the disorder, being reared by a schizophrenic parent will not likely cause that child to become a schizophrenic adult. Erlenmeyer-Kimling observes, "Parenthood and schizophrenia tend to mix poorly." She adds, "In addition to the genetic risks to the children of schizophrenic parents, there is considerable likelihood that any children of such parents will be exposed to a disrupted home environment, and frequently to a grossly unsuitable one. The birth of a child often exacerbates the patient's illness, and the responsibilities of bringing up the children tend to trigger further difficulties."

She holds out dim prospects for the happiness of the spouse of the schizophrenic and suggests that most counselors would probably agree that schizophrenic patients should be advised against parenthood, or, at least, adding to their families. She feels personally that they should probably be advised against marriage if they are unmarried at the time of counseling (p. 127).

Studies performed in several European countries prior to the extensive use of modern drug therapies and the resultant outpatient care for schizophrenic patients indicated reduced marriage and reproduction rates for that group in comparison to the rest of the population. Now, social changes and medical advances may change the marriage and reproductive probabilities for schizophrenics just as they have extended life expectancies and more nearly normalized living for victims of diabetes and hemophilia.

Erlenmeyer-Kimling presents the results of a study initiated by the late Franz Kallmann which investigated the impact of the trend toward shorter average hospital stays on the opportunity for schizophrenic patients to marry and become parents. The admission periods selected were 1934-36 and 1954-56, sampling consecutive admissions carried out in 11 New York State hospitals. The study showed a trend toward increased marital and reproductive rates for schizophrenic patients at the beginning of the period in which tranquilizing drugs and community-care programs became important treatment factors; the differential between the reproductive rates of schizophrenic patients and of the general population was declining; the largest differences between the two-era sample populations were observed among younger patients in the two groups (1976).

Early identification of persons predisposed to schizophrenia is one of the most important research goals at present, so that vulnerable individuals can be identified *before* they manifest flagrant symptoms. Early identification can increase the effectiveness of genetic counseling. Another benefit, of course, is the capability for preventive intervention in staving off, or at least moderating, the impact of a psychotic breakdown.

AFFECTIVE DISORDER

Affective disorder is easier to recognize and describe than is schizophrenia. On the face of it, then, it should be easier to diagnose and to treat, and it should be amenable to solid counseling. Both the treating physician and the genetic counselor need to ascertain the kind of depression that is being considered, bipolar (manic-depression) with its mood swings or unipolar (depression). They also need to know whether or not depression has occurred during a previous psychiatric or physical illness, which would not indicate a psychotic depressive pattern, or whether the episodes are sufficiently severe and recurrent to indicate that pattern.

Bipolar Affective Disorder

Some studies of bipolar or manic-depressive disease have found male bipolar patients who had no sign of affective illness on the maternal side but who showed only father-son transmission (Cadoret 1976); other studies have found high rates of mother-son transmission (Winokur and Clayton 1967). Cadoret admits (p. 118) that, aside from knowing whether there is father-to-son or mother-to-son transmission, any information that might be given to bipolar families seeking genetic counseling might be so tentative as to be useless.

Again, probabilities for risk cannot be assessed with certainty unless only one clear and specific type of genetic transmission is evidenced, as in the case of Huntington's chorea. In the case of bipolar depression, diagnosis is less clear-cut and transmission lines hence more crowded. If a relative of a patient with that illness is suffering from a mental or emotional problem, the diagnosis may more likely be bipolar psychosis. For example, Cadoret describes a bipolar female patient whose father had been misdiagnosed as having organic brain dysfunction. Knowledge of the daughter's manic-depression triggered reevaluation of the father's diagnosis. He responded well subsequent to antidepressant treatment. This clinician then comments that the application of a genetic principle led to more effective treatment for a sick family member (p. 119).

Genetic counseling for patients who themselves suffer bouts of manic-depression and who are concerned about the heritability for their children can only rarely include an approximate risk prediction. At the most, if the X-linked dominant transmission is assumed, the risk will be 50 percent. Further, the illness can usually be treated successfully and the patient does not suffer severe chronic handicaps. Thus, the outlook in counseling for manic-depression may not be so bleak as for schizophrenia.

Depressive Disease

The large proportion of psychotic diseases which may be diagnosed as unipolar depression suggests that Winokur et al. are correct in believing that subgrouping this large, amorphous group of people is advisable. From their series of studies, these researchers have evolved three groups which may be distinguished by specific family background. There is "depression spectrum disease," recognized as simple depression in a person who has a first-degree family member with antisocial personality or, more usually, alcoholism. "Pure depressive disease" denotes a patient with a depressed first-degree family member but no mania or alcoholism. Their third category, which they call "sporadic depressive disease," indicates an individual with no family history of depression or alcoholism among first-degree family members.

The researchers observed noticeable differences among the three groups in age of onset; those diagnosed as having sporadic depressive disease were considerably older when first admitted and at times of onset of the illness. Depression spectrum disease patients were likely to have suffered fewer episodes prior to admission to hospital and were likely to have suffered fewer depressive episodes and hospitalizations than the pure depressives. To these researchers, this indicates that a depression associated with alcoholism has fewer episodes and is apparently less severe when compared to a depressive illness associated with depression only in a family (Winokur et al. 1978).

Elsewhere, some of these same scientists write (Schlesser, Winokur, and Sherman 1979), "We suggest that unipolar primary depressive illness is three or more separate illnesses, each with a potentially distinctive mode of inheritance, pathophysiology, neurochemistry, clinical course, and treatment response."

Biochemical tests that can refine the diagnoses further may come; metabolic reactions may be determined. Meantime, application of refined methods of probability for risk assessment among relatives, present and potential, of those who are afflicted with unipolar depression is unlikely. Specific transmission has yet to be demonstrated. Physicians who are diagnosing or counseling depressive patients cannot always be certain that they have received an accurate family picture.

Problems and Possibilities

In summing up the problems of counseling on the affective disorders, Remi Cadoret remarks first on the usefulness of being able to distinguish between early and late onset of unipolar illness. This is helpful for predicting not only incidence of depression but also of other conditions such as alcoholism and perhaps sociopathy among relatives of early-onset patients. This prediction, quite importantly, "can lead to help in understanding the patient and his environment, which may be considerably disturbed by the presence of other ill relatives."

He points out, too, that it is important in predicting outcome for families to determine whether that family is unipolar or bipolar since the risk to relatives is lower in unipolar than in bipolar. The patient in the midst of a manic episode is easy to diagnose. A depressed patient is more difficult; the distinction sometimes is uncertain because manic episodes may have been overlooked (Cadoret 1976, p. 121). Those who are charged with identifying individuals at risk for developing some serious form of depression later in life will surely welcome further knowledge about the biochemistry of depression.

Steven Targum and Elliot Gershon mention the "perception of burden" which should be considered in the planning of families of persons having an affective disorder. They mention, for instance, the regrets which well spouses of bipolar patients have expressed concerning marriage and/or childbearing to persons who may be characterized, the authors say, by formation of dependent relationships which demand attention and unreciprocated love, by low tolerance for frustration, and by reliance on coercion, manipulation, pity, and constant "giving in" to their insatiable needs (in press). However, clinicians and counselors are able to say a great deal already through application of current knowledge about the genetic factors at work in the affective disorders.

Sex-Linked Chromosomal Disorders and Sociopathy

In the general population, criminality and other antisocial behavior can be attributed largely to nurture, the familial and social environment, as well as to nature, the genetic contribution. In similar environments, however, there is wide genetic variation among individuals. Accidents in chromosome distribution can place a person seriously at risk for sociopathic disorder in an environment which otherwise appears stable and respectable. Conversely, outstanding citizens have come from unpromising, even disreputable, milieus.

Chromosomal screening of men imprisoned for criminal behavior has revealed higher frequencies of sex chromosome abnormalities among them than among the general population. This has led to public debate and raised ethical issues concerning the legitimacy of "stigmatizing" people by identifying a chromosomal abnormality associated with sociopathies.

On the other hand, it has been suggested that sociopathies are also the way society labels behavior in one social class that is labeled medically in the middle class. Since prison inmates tend to come from among those who are less likely to use the medical system, most criminal aneuploids—those possessing a chromosome number that is not the usual multiple—are suffering from the same disease and physiological abnormality as hospital patients. They have gone "involuntarily to jail instead of voluntarily to hospitals," the difference between them being apparently that of education, upbringing, and society's labeling habits. Lawrence Razavi continues (1975, p. 82): ". . . all intersexuals in a sexually polarized society have some degree of sexual psychoneurosis or even psychosis; but only some, those who tend to be caught by the legal system, appear to solve their problems by excessive violence, perhaps because a certain amount of aggressiveness is necessary to deal with the limitations and stresses of life in their part of society. Experience with intersexual criminals suggests that the problem is not one of violence per se, but of a sexually immature emotional lability which goes along with a social readiness to use violence and an inescapable frequency of environmental stress that elicits fear and depression."

There may be XYYs and XXYs who go undetected and undiagnosed, whose lives are normal so far as the rest of the world may know. And some of the "no accounts" or "bad apples" of years past were quite probably undiagnosed and undetected XYYs and XXYs. The XO abnormality, or Turner's syndrome, whose overt symptoms are short stature and lack of functioning ovarian tissue which results in a lack of menstruation and breast development, can usually be recognized before birth by testing the amniotic cells. Some cases may be mosaics—or contain more than one genotype—so that diagnosis is difficult. In the case of a fetus diagnosed as Turner's syndrome, Omenn remarks (1976, p. 149) that a couple's decision about a pregnancy will be influenced by their "attitudes about abortion and their willingness to accept a 'less than perfect' child."

Pointing up the problem for both parents and counselors, Omenn writes (p. 148): "Should a family desperately want a baby or disapprove of abortion, the knowledge that the child is born with an XYY karyotype may interfere seriously with normal attitudes toward childbearing. It is obvious that moral, ethical, and legal problems arise from such situations especially when our knowledge of the natural history of the chromosomal syndrome is incomplete or derived from biased data. What do we tell the parents in such situations, when we are uncertain ourselves of the consequences of a particular test result?" In such difficult and sometimes heartrending cases, the genetic counselor may at least contribute as much information as we currently possess, to form a basis for the parents' decision.

SELF-FULFILLING PROPHECY OR BOON AND BLESSING?

Most people in our society have progressed past the belief that an afflicted child was sent as a punishment and past the days when the hastily averted gaze may have signaled judgment rather than sympathy. Our still-limited knowledge may have brought more understanding. It is freighted, however, with its own load of problems and decisions which family members must make, aided by genetic counselors.

Whether the counseling be prospective or retrospective, the clinician, in order to predict the course of the illness and the risk of recurrence, must make as complete and precise a diagnosis as possible. Parents of an affected child are anxious to know the risk to subsequent children, and families with affected relatives will need to know what to expect. In most cases, despite a detailed evaluation including radiologic studies, enzyme assays, and chromosome karyotyping, families can be provided with only a statistical statement.

A statistical statement may be confounded, though, in each individual instance. Illnesses take different courses, and even the most careful diagnosis is uncertain. Gilbert Omenn adds another possibility, that the affected child may have had a fresh mutation—the approximately 5-percent risk of retardation or birth defects that every couple takes when producing a child. He furnishes examples of probable statistical statements which can be provided to concerned families: The risk of recurrence in subsequent children is about 1 percent for most cases of Down's syndrome, about 5 percent for many birth defects, 25 percent for inborn errors of metabolism inherited as autosomal recessive disorders, or 50 percent for autosomal dominant conditions (1976, p. 142).

Prospective Counseling: Is the Burden Worth the Risk?

Ming Tsuang believes that a doctor must understand the counselees' intentions, intellectual capabilities, and states of mind before risk estimates can be presented effectively (1978, p.

1468). He asks, "Do they seek information, advice, or both? Do they want advice for their own offspring or someone else's? How sophisticated is their acquaintance with genetic facts and principles? What is their general level of intellect? Can they make rational decisions? What is their present emotional state? Finally, how do they perceive the risk/burden ratio in their situation?"

Burden is the "expected cost to a family of a recurrence of disorder." It is also disappointment over not bearing a "perfect" child; it is extra care, attention to diet, perhaps training and learning problems; and it is "baby sitting," possibly after the child's age peers are self-sufficient. Burden may include embarrassment, untold expense, grief, childlessness.

Once the burden and the risks are comprehended, the genetic counselor should explain the couple's options. Although the decision, whether abortion or longer-range family planning, belongs to the parents themselves, the doctor may carefully predict consequences of various courses of action and explain personal or professional preferences. After this, once the physician is sure the counselees understand not only the reasons but the potential results of their choice, the responsibility for a slow, careful decision is up to them.

A final stage of counseling is recommended by Tsuang. This is a followup which, conducted within a reasonable period of time, permits continued assessment of the counselees' understanding. Further, new data may allow for a more accurate revision of risk estimates.

What Are the Choices?

For a few extreme disorders, selective abortion of affected fetuses is sometimes chosen. Some parents prefer to prevent the birth of a child for whose life they can foresee only pain and for whom treatment would be unsatisfactory, perhaps useless. Among the conditions for which termination of pregnancy has been sought are the Down's syndrome, a number of rare inborn metabolic errors, and defects in neural-tube closure, a condition which can result in spina bifida, anencephaly, hydrocephalus, and other disorders. In the case of the Lesch-Nyhan syndrome, that rare X-linked recessive disorder which results in compulsive self-mutilation, detection *in utero* can suggest termination of a male fetus only since girls may be normal—or carriers.

There are problems. Amniocentesis, the procedure for obtaining amniotic fluid from the uterus with a small-guage hollow needle, depends heavily on the experience and skill of the obstetrician. Further, although the probability of obtaining a large enough sample for study is greater in more advanced stages of pregnancy, the procedure must occur early enough to allow 3 weeks for laboratory test results and for the safe, and legal, termination of the pregnancy, if that is indicated and subsequently chosen.

So far, there has not been a thorough evaluation of any possible risks of amniocentesis. Little, if any, trouble has been reported—no stories of maternal bleeding, induced miscarriages, and the like. But it is not yet known whether disturbing the volume or "ebb and flow" of the fluid could be deleterious.

"It would be a tragedy," writes Omenn, "if normal babies later suffered mild depression of IQ levels or some other subtle damage because of a diagnostic procedure aimed at detecting an abnormal fetus." He adds that, at the present, consideration of this possibility is prompting counseling advice that risk of the procedure may be sufficient to omit it in pregnancies with less than 1 percent recurrence risk for a given disease.

In any case, parents and counselors discuss the pros and cons of amniocentesis thoroughly, and parents are asked to sign a consent form indicating their understanding that there may be a low risk to mother and fetus; that more than one amniocentesis may be required to obtain enough fluid; that several precious weeks may pass before test results reveal that the cell cultures failed to grow; that the biochemical and chromosomal analyses may not be successful; that the *in vitro* results may not reflect the status of the fetus, especially in the case of twins; and that the possibility of birth defects and/or mental retardation from other causes cannot be eliminated by the specific tests given (Omenn 1976, p. 144).

Another significant issue is abortion, which is still disapproved by many people and is abhorrent to some. For many mothers the procedure seems simple, for others it can be physically and psychologically traumatic. Laws vary from place to place as to the basis for the procedure. Regardless of the laws on the books, abortion may continue to be viewed by many persons as an immoral affront, particularly if there is not absolute certainty about the mental and/or physical prospects of the fetus in question.

Is there a possibility that amniocentesis will be required for all pregnant women over a certain age, as the population grows ever larger and more unwieldy and the burdens of severely retarded or disturbed persons on Government and society are magnified? "Requiring" of this sort has never been a part of our national philosophy. Better understanding throughout the medical profession, transmitted to the public by education and counseling, may eventually solve this and other problems of prospective parents. Concerning prenatal testing, Gilbert Omenn writes (p. 155), "We should emphasize that amniocentesis is usually a life-saving procedure, reassuring a couple that the fetus is unaffected by a disease, the high risk for which might have led that couple to demand termination of all pregnancies."

A Few Alternatives Available

The choice of many families who knew they had a 25-percent chance of serious genetic disorder in a child has been to forego further pregnancy. Such a decision is no longer so obvious or advisable for those to whom prenatal diagnosis and abortion are relatively easy and acceptable to accomplish. For some autosomal recessive conditions—cystic fibrosis, occuring in 1/2,000 Caucasians; Tay-Sachs, 1/4,000 Ashkenazi Jews; and sickle-cell anemia, in 1/400 United States blacks—extensive screening among these populations, detection, and, ultimately, a massive educational program are recommended, especially for healthy young people who may be heterozygous carriers.

In the case of psychiatric disorders, screening and prenatal advice may be helpful to many families and may be socially useful in general. Because there is still so far to go, though, in biochemistry, neuroendocrinology, and other relevant disciplines before firm risk predictions can be made on individual bases, counselors in genetics can present only warnings, likelihoods, and sympathetic understanding. "We believe," write Targum and Gershon (in press), "that most self-referred consultands will be competent to make their own moral decision regarding marriage and childbearing, given the appropriate information and the unbiased support of the genetic psychiatric counselor. Perhaps those who most need this type of counsel will be too defensive or too incompetent to accept the information."

How Can Retrospective Counseling Help?

There have been many parents who have been relieved to know the real cause of their child's seeming backwardness, or aggressiveness, or awkwardness, or ungainly stature in comparison to other children. Many have been "glad it wasn't any worse," and others have sought compensatory help and advice so that neither child nor family suffered unduly.

The question of the "self-fulfilling prophecy" recurs when considering the behavioral disorders. If a child appears pensive, as everyone is entitled to be occasionally, might a parent assume he is evidencing early signs of grandpa's depression? Will parents from a family in which Huntington's chorea occurs see each twitch or behavioral problem as an early sign of this dreaded inheritance?

Considerations of this sort appear as an argument for any prenatal testing that is presently possible. It can be hoped that they will be seen, also, as an argument for further risk research and for well-informed genetic counseling that is both wise and cautious. If it is to be considered as a form of prevention, then the quality of information given to parents must be of the highest order commensurate with emotional and intellectual capacities.

When parents are informed of the increased risk for deviancy in XYY males, for instance, they should be reassured at the same time that the chances for institutionalization appear to be small and that their child is far from necessarily destined for a life of crime.

A portion of Razavi's plea for help and counseling for institutionalized persons with psychiatric and/or chromosome disorders can be more broadly applied to encompass the social concerns which might expand the viability of retrospective genetic counseling. He writes (1975, p. 91): "In handling these patients it seems most rational to integrate social background with the natural history of the patient's disorder in society at large, and thereby select a more precise training and education for each individual according to psychic and biological character ... the chief value of a genetical diagnosis is that it allows a taxonomical classification of sexual type, at an early stage in a troubled life, on which one may base the development of the patient personality according to social and educational opportunities."

Part II: Directions of Future Research in Behavioral Genetics

Eunice Corfman

Division of Scientific and Public Information
National Institute of Mental Health

As we have seen, genetic counseling even at its most knowledgeable, is a matter of laying out statistical risks based on twin, adoption, or family studies, and in a particular case, ascertaining the existence of the disorder or related ones in the immediate family of a patient or better, in a completely worked out genealogy. While such studies are significant in helping to establish in general the strength of the genetic component to the disorder, they do not help prediction for individuals, except by providing a general estimate of risk. Consequently, a challenge to clinical research of considerable magnitude and importance now is to discover within an individual the specific biogenic effects that contribute to psychiatric disorders. For only by such a narrowing down from patterns of transmission in populations to patterns within a given family to markers within a particular individual can prediction and counseling be made more specific. How is current research approaching this challenge? Several strategies are underway which are determining directions of research. We will be comparing one classic model of clinical research towards finding biological markers of psychiatric illness with a more recent genetic model. Strategies for this newer genetic model have been reviewed by Elliot

Gershon, Chief of NIMH's Intramural Psychogenetics Section. They include searching for markers of vulnerability through well-state studies and through challenge strategies; the application of these markers to pedigree studies; and the linkage of markers to single chromosomal locations.

Speaking generally, most strategies derive from an underlying and generating premise; different premises generate different kinds of strategies. An illustration may be helpful and clarifying. A classic, nongenetic model of psychiatric illness is built on the premise that such illness is essentially the result of stress in an otherwise normal person. On this premise, it is primarily the environmental and developmental influences, not genetic transmission, that bring about mental illness. For example, Meyersburg and Post have postulated that stress occurring during developmentally vulnerable periods may have effects on neural structures and function that produce dysfunction (1979). Strategies that derive from this premise include developing a description of the clinical psychophysiology of the illness compared to a normal state or a state of remission. Clues to the illness are sought in those physiological measures that are different in an ill patient compared to one in remission or a normal. Another related strategy on this model is to define the biological and clinical effects of the psychotropic drugs. Insofar as these drugs may be seen to mimic the effects of various illnesses, we may be able to infer, from the structural similarity of the drugs to endogenous substances or from their mechanisms of action, clues to the origins and processes of mental illness.

In contrast, the genetic premise, whose strategies we will review here, presupposes 1) some underlying vulnerability in predisposed people, whether or not they now have, have had, or have not had the illness; 2) genetic transmission as the essential test of etiology (Rieder and Gershon 1978; Gershon 1978).

The Search for Markers of Vulnerability

Strategies derived from this premise include *the search for markers of vulnerability*. One way is by means of *well-state studies*, of well relatives of patients or patients themselves in remission, who together with ill patients can be distinguished from normals by means of the marker. Compare this strategy, which seeks to distinguish genetic vulnerables, both ill and well, from normals who are not genetically vulnerable, to the classic model strategy, which seeks rather for markers to distinguish those in the well state from those who are ill. The strategies point up a fundamental difference as to which is the most relevant distinction. A second way to identify markers of vulnerability is by means of *pharmacologic challenges,* by which a selected drug is briefly used by vulnerables and normals to see if a differential response suggests a genetic variation (a difference in reaction attributable to genetic difference) and hence a useful marker of vulnerability.

Criteria for a *genetic* marker are:

1. The characteristic must be associated with an increased likelihood of the illness (although all people with the illness need not show the characteristic nor vice versa). It is then a marker for the illness, though not necessarily a genetic one.

2. It must be heritable and not be a secondary effect of the illness. That is, it must be genetic and not a result of having had the illness.

3. It must be observable (or evocable) in the well state in addition to the ill state. Since the marker is a predisposition to the illness, not of the illness itself, we should expect it in at least some well relatives and the recovered ill.

4. Transmission of both the characteristic and the illness must be related within pedigrees. This demonstration shows the characteristic is a necessary or contributing genetic factor in an illness.

A number of clinical studies using the classic model have compared characteristics of patients having episodes of illness with normals, to satisfy the first criterion of finding a characteristic for a marker that is associated with an increased likelihood of the illness. The genetic model directs a shift in focus from studying illness episodes themselves to studying the underlying biological vulnerability (or predisposition) present in remission as well as during episodes.

Some proposed markers have been enzymes of monoamine synthesis and metabolism, because some of the strongest hypotheses about the cause of psychotic disorders implicate the neurotransmitters, dopamine, norepinephrine, and serotonin. These enzymes are identifiable proteins, which make them especially attractive for genetic study because the protein structure is determined by a single genetic code sequence. Three amine-related enzymes of the central nervous system (CNS) have been given special attention as possible markers because they can be measured in peripheral blood. The enzymes are platelet monoamine oxidase (MAO), dopamine-beta-hydroxylase (DBH), and erythrodyte catechol-O-methyl transferase (COMT).

But the data from different studies do not show consistently that any of them is a reliable genetic marker.

Another strategy for locating a genetic marker for vulnerability is by pharmacologic challenge. This strategy deliberately administers a drug, in subclinical doses for a limited time, to patients in remission and to normal controls to see whether their reactions to the drug are different. The hypothesis is that a difference would mark a failure of normal adaptive response in an affected neurotransmitter system to this deliberately provoked challenge.

An example is the drug Arecoline, a cholinergic agonist—that is, a drug that occupies the same receptor sites as the neurotransmitter, acetylcholine. It triggers the onset of REM (rapid eye movement) sleep when infused during non-REM sleep in normal subjects. Some studies suggest that in patients with primary depression periods of REM sleep (when dreaming occurs) tend to occur earlier in the sleep cycle than in normals. On the hypothesis that *sensitivity* to REM induction by Arecoline continues with patients even in remission, so that their response to challenge will be different from that of normals, a recent study by Siteram, Nurnberger, Gershon, and Gillin has shown this to be the case in five manic-depressives in remission, but not in seven normal controls (Gershon et al. 1979). This suggests that a useful marker of vulnerability to manic-depression might be greater sensitivity of the cholinergic transmitter system. This finding will now be tested again, to see if it is replicated, and extended to pedigrees of such patients.

Application of Markers to Family Studies

A second kind of strategy, complementary to identifying markers of vulnerability, is the *application of such a marker to family studies* to see if the marker appears genetically transmitted within an identified set of relatives. Quite sophisticated statistical analyses are now possible with the help of computers.

Sometimes these analyses are of correlations between relatives, comparing many *pairs* of a proband (the initial subject with the illness) and a relative to look at correlation coefficients of incidence of the illness. But pair data, unlike whole family or pedigree data, do not lend themselves to the more powerful statistical analyses.

Segregation analysis uses an entire *sibship* (the set of all sibs) of the proband, and their parents, or a *whole pedigree,* or a large number of such units, to see the frequency of different kinds of relationships. Computers and their equation-solving capacity make possible control of these enormously increased numbers of relations, to derive segregation ratios (the proportion among progeny of alternate traits of a parental combination) and transmission patterns.

In all of these investigations the point is to trace within these family relations the linkage of some biological marker to the specified illness. Transmission of the illness within families must be correlated with transmission of the marker.

Linking the Marker to a Single Chromosome Location

If these family studies are promising, a third strategy is then to try to *link the marker to a single chromosomal location*. The linking of marker to chromosome may sometimes be done through biochemical evidence of the pathways linking genes to traits. In the absence of this evidence a link may be shown statistically by showing within a pedigree a linkage between a major gene, site of a neutral trait (the marker), and the illness. (Linkage should not be confused with association, which refers to correlated traits for whatever reason in the population as a whole—a more general concept and of less value in genetic analysis.) About 30 marker loci are now testable.

Human geneticists differ about the usefulness of linkage analysis for a condition for which no *major* locus is implicated. Linkage analysis is more suited to a single major locus model of genetic transmission than a multifactorial model, where the illness is determined at several loci, all of which can be partial contributors to it, along with environmental influences. But considerable advantages accrue if a linkage can be shown, among them the demonstration of genetic influence, identifying gene carriers in the general population, and specifying the mode of genetic transmission.

It is worth emphasizing that this type of linkage of illness to chromosome location is a different type of "genetic marker" from the linkage of the etiological marker to the illness. It differs in that the definition of the illness used to characterize the group being studied for chromosome marker may be entirely unrelated to the marker. Linkage to a chromosomal marker does not necessarily tell us anything about the cause of the disorder; it only, in effect, tells us *where* on the chromosome the genetic impact comes from.

These are some of the directions research is taking now. We do not yet have firm genetic markers for the psychotic illnesses in either of the two linkages described. But what we do have now are genetic strategies of considerable power, capable of definitive identification if and when the appropriate markers are found. These strategies are somewhat analogous to a new microscope possessing lenses with greater powers of resolution than heretofore available. The strategies are instruments that enlarge the field of potential findings. Psychiatric illnesses seem unlike many of the genetically simpler disorders attributable to single-gene enzyme defects (Rieder and Gershon 1978). One clinical illness may turn out to be composed of several biologic and genetic subtypes. And people who are biologically vulnerable or predisposed are not all ill; nor are all the ill demonstrably predisposed. These strategies must tease the genetic component from this complexity.

REFERENCES

Barchas, J.D.; Ciaranello, R.D.; Kessler, S.; and Hamburg, D.A. Genetic aspects of catecholamine synthesis. In: Fieve, R.R.; Rosenthal, D.; and Brill, H., eds. *Genetic Research in Psychiatry.* Baltimore: The Johns Hopkins University Press, 1975.

Bergler, J.H. Unpublished paper, 1979.

Brill, H. Presidential address: Nature and nurture as political issues. In: Fieve, R.R.; Rosenthal, D.; and Brill H., eds. *Genetic Research in Psychiatry.* Baltimore: The Johns Hopkins University Press, 1975.

Cadoret, R.J. The genetics of affective disorder and genetic counseling. *Social Biology,* 23(2):116–122, 1976.

Cromwell, R.L. Concluding comments: Genetic transmission. In: Wynne, L.C.; Cromwell, R.L.; and Matthysse, S., eds. *The Nature of Schizophrenia: New Approaches to Research and Treatment.* New York: Wiley, 1978.

Erlenmeyer-Kimling, L. Schizophrenia: A bag of dilemmas. *Social Biology,* 23(2):123–134, 1976.

Gershon, E.S. The search for genetic markers in affective disorders. In: Lipton, M.A.; DiMascio, A.; and Killam, K.F., eds. *Psychopharmacology: A Generation of Progress.* New York: Raven Press, 1978.

Gershon, E.S.; Nurnberger, J.N.; Sitaram, N.; and Gillin, J.C. Pharmacogenetics and the pharmacologic challenge strategy in clinical research: Studies of d-amphetamine and Arecoline. In: Saletu, B. et al. eds. Neuro-Psychopharmacology: Proceedings. New York: Pergamon, 1979.

Heston, L.L. Schizophrenia: Genetic factors. *Hospital Practice* 43–49, June 1977.

Kidd, K.K. A genetic perspective on schizophrenia. In: Wynne, L.C.; Cromwell, R.L.; and Matthysse, S., eds. *The Nature of Schizophrenia: New Approaches to Research and Treatment.* New York: Wiley, 1978.

Lubs, H.A. Frequency of genetic disease. In: Lubs, H.A., and de la Cruz, F., eds. *Genetic Counseling.* A Monograph of the National Institute of Child Health and Human Development. New York: Raven Press, 1977.

Meyersburg, H.A., and Post, R.M. An holistic developmental view of neural and psychological processes: A neurobiologic-psychoanalytic integration. *British Journal of Psychiatry,* 135:139, 155, 1979.

Omenn, G.A. Intrauterine diagnosis and genetic counseling in psychiatry. *Social Biology,* 23(2):132–157, 1976.

Rainer, J.D. Genetic knowledge and heredity counseling. New responsibilities for psychiatry. In: Fieve, R.R.; Rosenthal, D., and Brill, H., eds. *Genetic Research in Psychiatry.* Baltimore: The Johns Hopkins University Press, 1975.

Razavi, L. Cytogenetic and dermatoglyphic studies in sexual offenders, violent criminals, and aggressively behaved temporal lobe epileptics. In: Fieve, R.R.; Rosenthal, D., and Brill, H., eds. *Genetic Research in Psychiatry.* Baltimore: The Johns Hopkins University Press, 1975.

Rieder, R.O., and Gershon, E.S. Genetic strategies in biological psychiatry. *Archives of General Psychiatry,* 35:866–873, 1978.

Schlesser, M.A.; Winokur, G.; and Sherman, B.M. Genetic subtypes of unipolar primary depressive illness distinguished by hypothalamic-pituitary-adrenal axis activity. *The Lancet,* 1(8119):739–741, 1979.

Targum, S.D., and Gershon, E.S. Genetic counseling for affective illness. In: Belmaker, R.H., and van Praag, H.M., eds. *Mania: An Evolving Concept.* Holliswood, N.Y.: Spectrum, in press.

Tsuang, M.T. Genetic counseling for psychiatric patients and their families. *The American Journal of Psychiatry,* 135:12(1465–1475), 1978.

Winokur, G.; Behar, D.; Van Valkenburg, C.; and Lowry, M. Is a familial definition of depression both feasible and valid? *The Journal of Nervous and Mental Disease,* 166(11):764–768, 1978.

Winokur, G., and Clayton, P. Family history studies. I. Two types of affective disorders separated according to genetic and clinical factors. In: Wortes, J., ed. *Recent Advances in Biological Psychiatry,* Vol. IX. New York: Plenum, 1967.

APPENDIX

HEREDITY AND INTELLIGENCE

Correlations Between IQ Scores for Persons with Varying Degrees of Genetic Similarity

Identical twins reared together	.87
Identical twins reared apart	.75
Fraternal twins reared together	.53
Siblings reared together	.55
Parents and their children	.50
First cousins	.26
Grandparent-grandchild	.27
Unrelated children reared together	.23
Unrelated children reared apart	.00

Several studies have indicated that heredity plays a major role in determining intelligence. Hunt (1961), for example, found that the average correlation coefficient of intelligence test scores for identical twins is about .90. (For fraternal twins it is about .65). The study showed that as genetic similarity among individuals decreased, so did the similarity between intelligence test scores. Hence, identical twins had the highest correlation between test scores, while unrelated children had the lowest (.00). The correlations between the test scores in siblings was estimated at .50. Hunt's results clearly support a genetic influence.

Other studies involving twins also show heredity to be a major influence on intelligence. Muller (1925) found that identical twins showed negligible differences in intelligence test scores, even after they had been raised in different environments ("split-twin" studies). In a classic case involving identical twins, Jessie and Bessie, Muller found that Bessie, who grew up in a relatively poor family and had only four year of school, scored two points higher on an IQ test than Jessie, who grew up in an affluent family and had completed high school.

Newman, Freeman, and Holzinger (1937) substantiated Muller's findings. They found that monozygotic twins who were raised in disimilar environments obtained intelligence tests scores which were more similar than were the scores of dyzygotic (fraternal) twins who grew up together.

Arthur Jensen (1969) advanced a controversial theory which holds that heredity is most responsible for an individual's intelligence level. He estimated that the contribution of heredity to intelligence is 80 percent while the environment's contribution is only 20 percent. This estimate was derived from comparisons of intelligence test scores among different racial groups. Jensen found that American Blacks performed less well than American whites and that Caucasians performed less well than Orientals. Jensen argues that the most important environmental influences on intelligence are prenatal; however, differences in test scores between those of different racial and social classes cannot be attributed to environment alone. This conclusion invites racist interpretations and has been hotly debated. According to Jensen's 80-20 ratio of the influence of heredity and environment, it would follow that American whites have more potential for intellectual activities than American Blacks. Jensen also argued that compensatory educational programs are useless.

Urie Bronfenbrenner (1972) has questioned Jensen's 80-20 ratio. He suggests that the twin studies upon which Jensen bases his claims involved too many uncontrolled variables. For example, when Jensen studied the intelligence comparison of monozygotic twins who were raised in different environments, the cultural aspects were really quite similar.

Newman, Freeman, and Holzinger (1937) had previously reached an estimate of 50 percecnt for the amount of genetic influence on heredity. Later on, Fehr (1969) also found a 50 percent or less genetic influence on intelligence.

EUGENICS AND INTELLIGENCE

Eugenics can be defined as a system of genetic engineering whereby certain individuals are chosen for reproduction on the basis of their genetic traits. This practice of selective breeding was first advocated by Francis Galton, Charles Darwin's cousin. During the last half of the nineteenth century, he was concerned about the diminishing number of gifted men in England, and he believed that only parents who possessed favorable genetic characteristics should be allowed to bear children.

The practice of eugenics with animals is popular and widespread, particularly among farmers who wish to raise superior farm animals.

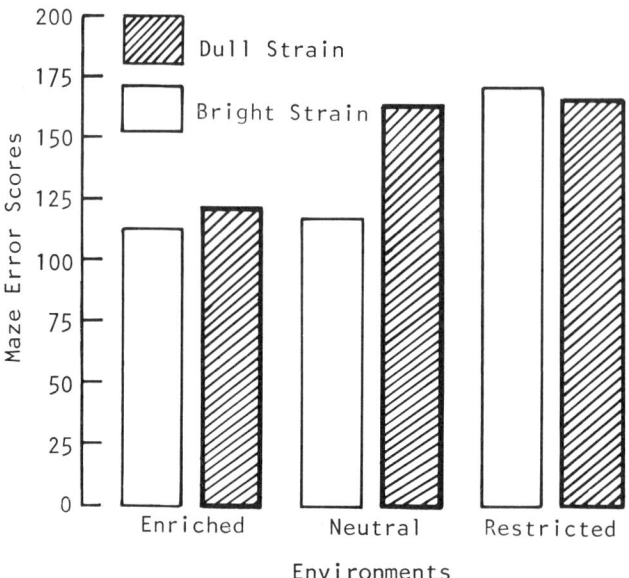

The effects of heredity and environment
on maze learning in rats.

An experiment conducted by Tryon (1940) at the University of California tested the inheritance of learning ability in rats. Tryon succeeded in producing a strain of bright rats and a strain of dull rats by means of selective breeding. Taking a group of rats, half male and half female, Tryon ran them through a maze and recorded the number of errors each rat made. He mated those the scores showed to be the bright rats and did the same with the dull rats. He subjected the offspring of both the bright rats and the dull rats to the same maze test, again recording their scores. He then mated the brightest offspring of the bright rats with each other and the dullest rats of the dull parents with each other. Tryon continued with this procedure--mating the brightest of the bright rats with each other and the dullest of the dull rats with each other--for a total of 18 generations. After 8 generations he found that the dullest rats from the bright group were brighter than the brightest rats from the dull group. Thus, Tryon showed that eugenics under controlled conditions can produce superior strains of animals and supports the theory that intelligence depends a great deal upon heredity.

Searle (1949), however, tested the same strains of rats and found that neither strain was universally bright or universally dull; that is, those rats who scored well on the maze-running task did not perform well on all types of tasks. In the same way, the dull rats who did poorly at maze-running were not poor at every task. However, those findings do not completely discredit Tryon's findings since Tryon was looking at maze-learning ability alone.

BIOLOGICAL VERSUS ENVIRONMENTAL DETERMINISM

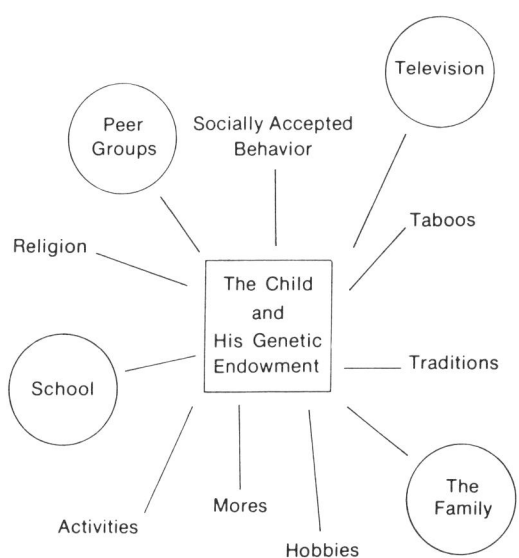

The child's culture interacts with his genetic endowment to determine the sort of adult he will become.

Biological determinism refers to the attitude that biological heredity is the only factor that determines differences among individuals. The British scientist Sir Francis Galton published a book in 1869 entitled <u>Hereditary Genius</u> in which he stated that man's abilities are derived from innate intelligence regardless of his surrounding environment. This was a popular view at the time. Parents and teachers were encouraged not to interfere with the natural development of their children. Children should be safeguarded from the harmful effects of the environment rather than be stimulated by it. In other words environmental factors could only hurt the child's development, never help it.

Environmental determinism stresses the importance of cultural influences and other aspects of the environment that influence human development. Theorists of this viewpoint believe that human development can be controlled by manipulating the environment. Today, most "environmentalists" do not deny the importance of genetics in development but sometimes tend to ignore it in their research and writing.

The debate between biological determinists and environmental determinists is still taking place. It is popularly referred to as the nature--nurture controversy.

Most psychologists today agree that both heredity and environment, or nature and nurture, interact to determine behavior. Nature, inheritance, sets the potentialities for behavior, while nurture, the environment, determines the extent to which these potentialities will be realized. For example, an individual may be born with a certain potential for intelligence, but whether this potential is fulfilled depends on various factors in the environment: availability of books and schools, encouragement of learning in the home, etc.

Genotype refers to the genetic constitution of an organism; its potential, in other words phenotype, refers to the observable characteristics of an organism; that is; the extent to which its potential is realized. The genotype, nature, does not set fixed genetic limits for behavior. Because the genotype must act through the environment to produce a particular behavior, no such limits exist. The important point is that behavior is not solely dependent upon an individual's genotype, but rather upon the interaction of the genotype with the individual's environment.

The epigenetic approach to the study of behavior development is one which concentrates on the interaction of nature and nurture. The conceptual basis of this approach is that there is an on-going interplay, during the course of development, between behaviors that are genetically controlled and the environment. The result of this interplay is the observed behavior--the phenotype.

One study that illustrates the epigenetic approach is one concerning the accuracy of pecking behavior of seagulls in Eastern North America. Hungry chicks ordinarily peck at the red bills of mother birds in order to make them regurgitate food. Using a controlled laboratory setting, cards on which the mother gull head was drawn were presented to young chicks. Records were kept of the number of times the chicks pecked at the red bill drawn on the card. It was found that on the first day after hatching, approximately 33 percent of the pecks were hits while on the fourth day, after much practice, about 80 percent of the pecks were hits. Here we see that genetics provides the basis or potential for pecking while environmental experience determines the extent to which this potential will be fulfilled. Thus we see the epigenetic approach taken toward this study--the interaction of nature and nurture.

Stern's widely accepted hypothesis on the contribution of heredity and environment with regard to behavior development has been termed the "rubber-band" hypothesis. In it, he models by analogy a genetic trait to a rubber band, and the extent to which the trait develops, to the length to which the rubber band stretches. That is, an individual is born with a certain amount of genetic endowment for a particular trait, and the amount the trait develops depends on the environment. Stern suggests that "different people initially may have been given different lengths of unstretched endowment, but the natural forces of the environment may have stretched their expression to equal length, or led to differences in attained length sometimes corresponding to their innate differences and at other times in reverse of the relation."

BEHAVIOR DISORDERS AND GENETICS

Several types of behavior disorders can be inherited through the genes. Brain damage may be inherited when the body, as a result of genetic inheritance, is not able to produce enough of the enzymes needed for proper development.

Hemiplegia and some types of cerebral palsy can be genetically transmitted. Hemiplegia is a condition in which one side of the body is paralyzed. Cerebral palsy is a disorder in which the brain cannot control muscle movement.

Learning disorders and various types of retardation can be traced to genetic causes. Down's Syndrome, or mongolism, is one such form of retardation. This condition is most often found in children of very young mothers or mothers who are over forty years of age. It is caused when the paired chromosomes in an egg or sperm cell fail to separate properly when they join to form a zygote. A mongoloid child is characterized by a low intelligence level, an unusual skin fold in the corners of the eyes, a wide nose, and protruding tongue. A child with Down's Syndrome has an average life span of ten years.

With regard to mental illness, Gottesman and Shields (1966) found a higher concordance between identical twins who had schizophrenia than between fraternal twins with the same disorder. This means that of all twin children studied who had schizophrenia, more individuals of identical pairs had a schizophrenic twin than the fraternal members. Gottesman and Shields found the concordance rate for identical twins--that is, the percentage of those twins in which both members are schizophrenic--to be 42 percent, based on a sample of 28 pairs of identical twins. They found the concordance rate for fraternal twins to be only 9 percent, based on a sample of 34 pairs of twins.

Schizophrenia is one of the most commonly diagnosed mental illnesses in the United States. It is characterized by the individual's detachment and inability to relate to the surrounding environment. Some studies have indicated that schizophrenia can be, in part, genetically inherited. But most research cites stress in family life and abnormalities in the nervous system as the major (or at least precipitating) causes.

Concordance is the similarity in psychiatric diagnosis or other traits in a pair of twins. To research the role of genetic factors in schizophrenia, twin studies were designed to find out whether the concordance rate for schizophrenia is greater for identical (monozygotic) twins, than it is for fraternal (dizygotic) twins.

In a major study in Norway, Kringlen (1967) found a 38 percent concordance for identical twins as contrasted with 10 percent in fraternal twins. There were 55 pairs of identical twins and 172 pairs of fraternal twins. Gottesman and Shields (1972) found a concordance rate of 42 percent for identical (MZ) and 9 percent for fraternal twins (DZ) who were hospitalized and diagnosed as schizophrenic. They also found that the concordance was much higher for twins with severe schizophrenic disorders than for those with mild symptoms.

Although the concordance rate for schizophrenia in identical twins is high, the discordance rate is higher. If schizophrenia were solely the result of genetic factors, the concordance rate for identical twins would be one hundred. One must also take into account that since the twins have been reared together, a common environment rather than common genetic factors may account for concordance rates.

In a study by Rosenthal (1970) 16 pairs of monozygotic twins were reared apart from very early childhood. Of the 16 pairs, 10 were concordant and 6 were discordant. The concordance rate of this limited sample was 62.5 percent, a finding that supports the view that a predisposition for schizophrenia is genetically transmitted. Because of the small sample size, however, the data cannot be regarded as conclusive.

Although studies have not proven that schizophrenia is solely transmitted through genetic factors, the findings support the view that a predisposition for schizophrenia may exist. Here it is presumed that certain individuals are more prone to develop schizophrenia if placed under severe stress. However, given a more favorable life situation, the individuals inherent vulnerability may never exhibit itself in the form of schizophrenic behavior.

ENVIRONMENT AND INTELLIGENCE

It is clear to most psychologists that intelligence is determined by both genetic and environmental factors. The predominant view is that the genetic factors place an absolute ceiling on an individual's intelligence potential, and he may or may not reach this potential, depending on how enriched or deprived his environment is. Thus an individual may have the potential to develop an IQ of 120, but if his learning environment was bad in childhood his IQ may never develop past 105. It is believed that there is a critical period in a person's childhood in which he must develop certain intellectual functions. If this critical period is characterized by emotional, intellectual, social, or physical (e.g., food, shelter) deprivation, then the person's IQ will not develop to its full potential. This is usually an irreversible result; that is, even if the person's later life is enriched, he will not be able to raise his IQ significantly and reach his potential.

There are many instances in which a young child's IQ increases with an improvement in environment. Skeels (1966) conducted a study in which he took very young, girls (under 19.3 months) from an orphanage in which the environment was seriously depressed and placed them in an institution for the mentally retarded. Their IQ's improved from a mean of 65 for the group to a mean of 91.8 during an 18 month period. The cause of the subnormal mean IQ of 65 was the lack of personal attention in the orphanage. The girls had a severe emotional deficit which impaired their intellectual development. In the institution for the mentally retarded, the girls experienced improved caretaking by retarded adult women who had a mental age of five to nine years. These retarded women were attentive and affectionate towards the little girls and this compensated for the previous emotional deficit. Eleven of the thirteen children placed in the institution for the mentally retarded were adopted. Another group of 12 little girls stayed in the orphanage; their mean IQ dropped from 86.7 to 60.5 in two years. None of these girls was adopted, and one of them died.

In another study by Newman et al (1937), it was found that the average correlation of intelligence between identical twins raised in very similar environments was .88; whereas the correlation of intelligence for fraternal twins raised in very similar environments was only .63. This difference between identical and fraternal twins indicates that heredity is an important factor in determining IQ. However, the correlation of intelligence for siblings is only .51-.53; and since any two siblings are as genetically similar as any two fraternal twins the difference in the correlation between the two groups must be attributed to environmental factors. Obviously, fraternal twins are raised in more similar environments than are siblings in general. When the correlation of intelligence for identical twins raised apart from each other was computed it was found that the correlation was .77. The correlation of intelligence for identical twins raised apart was still higher than the correlation of intelligence for fraternal twins and for siblings. This evidence seems to suggest that heredity plays a more important role in determining intelligence than does environment. However, this study was conducted with a small number of subjects and therefore the results are not as impressive or authoritative as they seem.

Thus it can be seen that heredity and environment combine to determine an individual's intelligence. It appears that a child's father's occupation is the most accurate predictor of a child's future IQ. Father's occupation is even more effective in prediction than father's IQ. Psychologists are still actively involved in determining heritability indices for intelligence. A heritability index represents what proportion of the variance of a particular trait in a given population (the individual differences--that is, why person A has an IQ of 80 and person B has an IQ of 150) is attributable to hereditary or genetic factors. The higher the heritability index, the more pessimistic is the outlook for raising an individual's IQ.

Several studies have indicated that children who are raised in isolated regions, where they are deprived of the degree or quality of environmental stimulation which other children in less isolated areas receive, are intellectually inferior. One famous study concerning this aspect was done by Sherman and Key (1932) on a group of isolated mountain children which they call the "Hollow Children."

The Hollow Children lived in various isolated villages in the Blue Ridge Mountains about 180 miles west of Washington D.C. Interestingly, these villages had remained isolated in varying degrees for decades prior to the study. Sherman and Key chose four villages in the region to study. Each varied in its amount of contact with the outside world. For example, Colvin was the innermost village and was the most isolated. The other villages, in order of most isolation to least, were Needles, Rigby, and Oakton. Each of these villages varied in the amount of schooling available. The researchers found that Colvin provided the least amount of schooling. During the 12 years previous to the study, a school had been open a total of only 16 months. It was found that only three of the adult residents were literate. In Rigby, which was closer to civilization, a school had been open for 66.5 months during the same amount of time. Many of the adults here were literate and there was a post office which provided some contact with civilization. A village called Briarsville which was located at the foot of the mountain was much less isolated than the mountain villages and was used as a control group for the study.

The researchers used various intelligence tests such as the Stanford-Binet and the Goodenough Draw-A-Man test, to compare the intellectual ability of the extremely isolated children in Colvin to the less isolated children in Briarsville. Sherman and Key wanted to see if the environmental deprivation had any effect on intellectual development. Since most of the residents of the Blue Mountain area were of common ancestry (English and Scotch-Irish), it was assumed that the genetic pool was similar for most of the people in the region. Since heredity as a factor in intellectual development was more or less in check (controled for), the researchers could more accurately gauge environmental influence as a factor.

Two significant findings came out of this study. The first was that scores on the intelligence tests increased as one moved from the most isolated village to the least isolated. Hence, the children of Colvin had the lowest scores while the children of Briarsville, the control group, had the highest. Secondly, it was found that older children (14 and 15-year-olds) performed worse than younger children. The very young children were not significantly below the test norms. This would indicate that a deprived environment has a detrimental effect on intellectual development. Both of these findings suggest that an individual's environment has an important influence on his ability to do well on these kinds of tests.

As a cautionary note: there is no accurate definition of intelligence. In this study, it was defined as that which intelligence tests measure.

One of Sherman and Key's most important and basic conclusions was that "children develop only as the environment demands development."

David Krech and his associates (1960; 1962; 1966) in their experiments with rats found that rearing in an enriched environment produces a higher level of intelligence than rearing which takes place in a relatively impoverished environment.

In their experiments, Krech and his associates created two types of environments for the rats: enriched and deprived. A rich environment for the rats was a large, well-lighted cage that had plenty of toys to keep them occupied (e.g., marbles, exercise wheels and gnawing or scratching posts). A deprived environment was a small, dimly lit cage with none of the toys of the enriched cage. Furthermore, the deprived rat's cage was lined with tin in order to obscure his view. As opposed to the enriched rat, the deprived rat was permitted no contact with humans or other rats.

Krech found that the rats raised in the enriched environment possessed a higher intelligence level (as measured by maze learning ability) compared to the rats who were raised in the impoverished environment. He also found that the chemical composition of the brain of a rat raised in an enriched environment differed significantly from that of a rat reared in a deprived environment. The rats from the enriched environments were found to have more cholinesterase and acetylcholinesterase in their brains than the deprived rats. These two chemicals are believed to affect the transmission of impulses from one brain neuron to another. An increase in the amount of these chemicals in the brain would facilitate learning ability.

Enriched rats also have a larger visual cortex than deprived rats, which would also aid learning. In addition, enriched rats developed brains which were heavier than those of deprived rats.

Krech's studies clearly suggest that an enriched environment is beneficial to a rat's development, and an impoverished one is detrimental. There is every reason to believe that we can safely assume that this conclusion is also true for humans; however, there is a dearth of evidence to support this conclusion.

Research has indicated that emotional responsivity can be inherited in animals. Significant differences can be seen in the degree of savageness between wild and domesticated animals. Wild animals such as gray rats, wolves, and lions are innately savage whereas their domesticated counterparts--white rats, dogs, and cats are innately tame. Even if these animals' environments are changed, they still retain their respective potentials for tameness and savageness. A wild animal raised by humans from birth will always remain potentially savage and will never become as tame as a domesticated animal. A domesticated animal released to the wild may become relativity savage, but will remain more "tamable" than animals whose ancestors were wild.

Several experiments have been conducted to detect emotional inheritance. In one study, Hall (1938) tested the level of emotionality in 145 rats by using a device called an "open field"--a large compartment that usually elicits fear. Of these rats, Hall mated 7 of the most emotional females with the 7 most emotional males. Likewise, the 7 least emotional females were mated with the 7 least emotional males. This inbreeding continued for several generations. It was found that the offspring of the emotional rats were a great deal more emotional than the offspring of the unemotional rats. This difference was considerable, even in the first generation of offspring. In Hall's test of emotional responsivity, he found that the scores of the emotional offspring were seven times those of the unemotional offspring.

The inheritance of emotions has also been studied in humans, although not as extensively as in animals because of practical and ethical considerations. In one experiment, Jost and Sontag (1944) studied children between 6 and 12 years of age over a period of 3 years. The researchers measured various bodily states which are known to at least partly reflect emotion. Among these were skin resistance, pulse and respiration rates, and salivation. For subjects, the researchers chose pairs of identical twins, pairs of siblings, and pairs of unrelated children. They wanted to see whether twins (who have identical heredity) are more similar in emotional responsitivity--as reflected in bodily states associated with it--than siblings and unrelated children. Using a correlation coefficient (a number between + 1.00 and - 1.00 used to express the degree of relationship between two sets of measurements arranged in pairs) the researchers combined the correlations of the scores on the different measures of bodily states. It was found that the correlations between identical twins were consistently higher than those for siblings, and the correlations for siblings were higher than those for unrelated children. Overall, the relationships clearly showed that heredity is influential in bodily states associated with emotion.

Twin studies have been used to determine the amount of genetic influence on personality. Rosanoff, Handy, and Plesset (1937) found a case in which a pair of identical twins who grew up in different families were both found to be sexually promiscuous. Both girls entered institutions at the age of twenty.

Lindeman (1969) also found marked similarities between monozygotic twins raised in different environments. In a case of two brothers, one was raised in a secure and stable home while the other was moved from one foster home to another. As adults, they were found to be very similar with regard to attitudes, work and even gestures. They even volunteered for military service within eight days of each other.

A study by Eysenck (1964) indicated that emotional disorders can be caused by genetic influence. It was found that with regard to criminal behavior and alcoholism, identical twins raised in different environments are more likely to display similar tendencies than fraternal twins. However, other studies of monozygotic twins who were raised in diverse environments show that behavior and personality traits differ, thus favoring emphasis on environmental influence. Some studies have even indicated that monozygotic twins who were raised in the same environment often develop differing personality traits. A study done by Shields (1962), for example, reported that in one case study one of the twins (MZ) took a physically dominant role with the other and became a leader.

CULTURE AND INTELLIGENCE

Culture is a difficult term to define; it encompasses much. We can say that a culture is the sum total of the mores, taboos, traditions, beliefs, value systems, and implicit and explicit standards of behavior that characterize a particular society. A child's culture interacts with his genetic endowment in determining the sort of individual he will become.

There are several means by which a child comes to learn about his culture and the ways in which he is expected to behave. Among these are school, religious institutions, the family, peer groups, and the mass media.

It is cultural characteristics that produce the greatest differences among various races of people, since genetically determined characteristics are actually very similar. People possess more genetically determined characteristics that are similar (i.e., two arms, two legs, a single head) than characteristics that differ (i.e., skin pigmentation, hair texture). It is the environmental, cultural variations which are most outstanding: language, social customs, attire. It is probably the case that a major portion, if not most, of an individual's behavior is determined by the particular culture in which he is raised.

Benedict (1934) believes that human behavior can only be understood and evaluated within the context of its cultural environment. He calls this belief "cultural relativity." Most behavioral scientists today would tend to agree with this viewpoint. Based on cross-cultural research, it is quite clear that certain behaviors which are acceptable in one society may not be acceptable in another. For example, the cannabalism of certain secluded Amazon tribes would not constitute acceptable behavior in the United States, but it is the norm in these South American tribes.

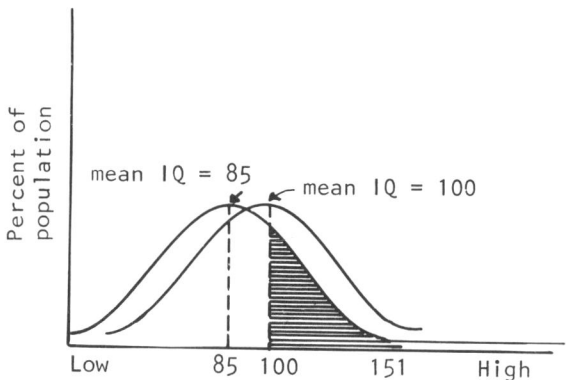

There is no good evidence that there are cultural differences in intelligence. There are, however, mean IQ differences between subcultural groups. Since IQ tests are designed for use in Western cultures, and there are no good cross-cultural (or cross-subcultural) intelligence tests, it is difficult to make valid comparisons among various cultural and subcultural groups. It seems reasonable that there are really no cultural differences in innate intelligence. Certainly if intelligence is the ability of an individual to deal effectively with his environment, there are no cultural differences in intelligence. People in other cultures deal as effectively with their environments as do people in Western culture. In fact, there is a considerably lower incidence of mental illness in non-Western cultures. (Mental illness is often used as an index for maladjustment to a particular society.)

86
88